SURVEYING

Godwin Study Guides

Advanced Theory of Structures
Design of Reinforced Concrete Elements
Fluid Mechanics
Materials and Structures
Mathematics for Engineers
Soil Mechanics
Specifications and Quantities
Structural Steelwork
Theory of Structures

GODWIN STUDY GUIDES

SURVEYING

R. H. DUGDALE

B.Sc.(Eng), M.Sc., F.I.O.B., C.Eng. M.I.C.E., A.M.B.I.M., A.C.G.I.
Head of Department of Construction & Surveying
Erith College of Technology

THIRD EDITION

GEORGE GODWIN LIMITED

The book publishing subsidiary
of The Builder Group

First published 1966 by Macdonald & Evans Ltd
in the *Examination Subjects for Engineers and Builders* series
Second edition 1970. Reprinted 1974
Third edition published 1980 by George Godwin Ltd
as a *Godwin Study Guide*

© George Godwin 1980

George Godwin Limited
The book publishing subsidiary
of The Builder Group
1–3 Pemberton Row
Red Lion Court
Fleet Street
London EC4

British Library Cataloguing in Publication Data

Dugdale, Roger Houghton
  Surveying – 3rd ed. – (Godwin Study Guides).
  1. Surveying
  I. Title
  526.9   TA545
  ISBN 0 7114 5641 0

Cover photograph by courtesy of
Wyld Heerbruk UK Ltd

Filmset by Northumberland Press Ltd, Gateshead, Tyne and Wear
Printed in Great Britain by Fletcher & Son Ltd, Norwich

GENERAL INTRODUCTION

THIS series was originally designed as an aid to students studying for technical examinations, the aim of each book being to provide a clear concise guide to the *basic principles* of the subject, reinforced by worked examples carefully selected to illustrate the text. The success of the series with students has justified the original aim, but it became apparent that qualified professional engineers in mid-career were also finding the books useful.

In recognition of this need, the books in the series have been enlarged to cover a wider range of topics, whilst maintaining the concise form of presentation.

It is our belief that this increase in content should help students to see their study material in a more practical context without detracting from the value of the book as an aid to passing examinations. Equally, it is believed that the additional material will present a more complete picture to professional engineers of topics which they have not had occasion to use since completing their original studies.

A list of other books in the series is given at the front of this book. Further details may be obtained from the publishers.

M. J. Smith
General Editor

PREFACE TO THE THIRD EDITION

The object of this book is to acquaint the reader with the principles of surveying in as short a space and as simple a manner as possible. For the most part these principles have not changed since the last major revision although progress in technology, particularly electronics, has been reflected in the development of electromagnetic distance measurement (i.e. E.D.M.), photogrammetry and, to a lesser extent, theodolites and trigonometrical levelling. New subject-matter has, therefore, been included on these topics and also on mass haul diagrams and error theory.

In the text, references are still made to the use of logarithms where it is thought that this will be helpful in understanding calculation processes but, where this occurs, it should not be interpreted as recommending that logarithms are preferred to calculators as a computational aid. One of the most significant advances in electronics has been the rapid development, and fall in prices, of pocket calculators and these are now not only quicker but in most cases more accurate than logarithms. Every surveying student who uses this book is urged to get one. At the very least it should be capable of handling, in addition to the arithmetical processes, trig functions and log functions. Preferably it should also be programmable and equipped with what Texas Instruments call the A.O.S. entry system in which $\times$, $\div$, $+$ and $-$ are automatically dealt with in the accepted order.

The many new examples, which have been added throughout the book, are derived from the following sources:

G.C.E. "A" Level Surveying
Higher National Certificate examinations in Civil Engineering
Higher National Diploma examinations in Civil Engineering
First-year degree examinations in Civil Engineering
Final-year degree examinations in Civil Engineering
R.I.C.S. professional examinations and
I.Q.S. professional examinations

Clearly, the same basic principles of surveying are encountered in a wide variety of examinations, for which this book will also be suitable.

The author would like to thank the appropriate bodies for permission to use questions from their examination papers and these include: the Associated Examination Board, the Institute of Quantity

Surveyors, the Royal Institution of Chartered Surveyors and the School of Civil Engineering at Thames Polytechnic.

October 1979   R.H.D.

CONTENTS

CHAPTER 1

SURVEYING INSTRUMENTS

LINEAR MEASUREMENT

Equipment available

The following equipment, in ascending order of accuracy, is available for linear measurements.

(a) *Chain.* Usually 20 m, 25 m, 30 m or 50 m long. A chain can be read direct to the nearest link (200 mm in length), every tenth link being marked by a tally.

(b) *Linen tapes.* May or may not be plastic coated, sometimes reinforced by a metallic thread, these are available in lengths of 10 m, 15 m, 20 m, 25 m or 30 m. They are usually graduated at intervals of either 5 mm or 10 mm dependent upon the inherent accuracy of the tape.

(c) *Steel tapes and bands.* These are commonly available in various lengths up to 100 m and can often be read direct to the nearest millimetre. For the highest degree of accuracy steel bands can be supported at each end on tripods on one of which is mounted a travelling microscope for reading the band against a fine line, engraved on the tripod head, exactly above the station.

Maps and plans always show horizontal distances. Consequently measurements made by chain, tape or band between stations at different altitudes must be corrected for slope. In "base-line measurement", where a high degree of accuracy is required, a steel band (possibly a steel tape) is used at a known tension and its temperature is measured. Corrections are then made for "sag", "stretch", "expansion", etc., as shown in Chapter 2.

Chain surveying

Chain surveying is by far the commonest surveying operation to involve linear measurements. A complete description would fill a chapter, for this heading is usually taken to include subsidiary operations, such as ranging and setting out by chain or tape, in addition to topographical surveying, which is briefly described below. A fuller account of chain surveying is given in, for example, Wilson's *Land Surveying* (Macdonald & Evans).

The essential equipment in chain surveying consists of a chain, tape, ranging rods, arrows and field book. To survey an area, station points, marked by pegs, are so placed that:

(a) they are intervisible;

(b) the chain lines between stations form a self-checking network of triangles. Moreover, the triangles should be "well-conditioned"— a well-conditioned triangle being one whose angles are all greater than about 45°;

(c) the chain-lines pass near the most important features of the area.

Having fixed the stations, each line is chained in turn, the procedure being the same for each.

(a) If it exceeds the chain in length the chain-line is first marked out by the brightly coloured ranging rods which are carefully lined up, by eye, between the stations each end of the line. This operation is called ranging.

(b) The distance between the two stations is then carefully measured by chain, each chain length being marked by an arrow, a pointed straight steel wire usually about 300 mm to 450 mm long. Each time the chain is laid down, offsets (i.e. measurements at right-angles to the chain-line) are taken to the important features. On one line, for example, the corner of a building may make an offset of 7 m at chainage 69 m, chainage and offset thus providing a system of co-ordinates from which a plan may be drawn. These measurements are booked a recognised way in a standardised book called a field book.

The minimum team for a chain survey is two men, for it takes two to pull the chain taut each time it is laid down. Three men, a surveyor and two chainmen, is, however, a more convenient number.

The right-angle for short offsets is usually judged by eye; for accurate long offsets, various hand instruments such as the optical square or cross staff may be used to advantage. Alternatively, the properties of the 3,4,5 (Pythagoras) triangle may be employed to set out a right angle, by judicious use of the tape.

LEVEL READINGS AND ANGLE MEASUREMENTS

Level readings and angle measurements are probably more common than linear measurements in engineering surveying. The instruments used for these observations are, respectively, levels and theodolites, which have a common feature in the telescope.

The telescope

In its simplest form, the telescope comprises an objective, an eyepiece and a diaphragm. The objective produces an inverted image at the diaphragm, and this is magnified by the eyepiece as shown in Fig. 1 (*a*). The telescope has two main functions.

(*a*) It fixes accurately the line of sight (also called the line of collimation) from a point over the instrument station to some distant point. The line of sight is the straight line joining the centre of the objective and the intersection point of the cross hairs.

(*b*) Provided that its diaphragm has stadia lines in addition to cross hairs (*see* Fig. 1 (*b*)), it can be used to measure the distance between the two points, as in tacheometry. Then A_1 and B_1 are the staff readings given by the stadia lines A_2 and B_2, and A_3B_3 is the image of A_2B_2 magnified by the eyepiece.

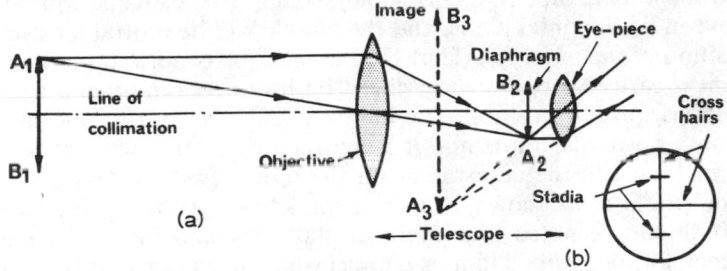

FIG. 1. (*a*) *Simplified optical arrangement of the telescope.* (*b*) *View through the eyepiece of the telescope.*

The telescope is said to be correctly focused when (i) a sharp image is obtained and (ii) there is no parallax. Parallax is tested for by moving the eye across the eyepiece; if, then, the image appears to move relative to the cross hairs the two are said to exhibit parallax. There are, therefore, two focusing operations which have to be performed:

(*a*) to focus the cross hairs by screwing or sliding the eyepiece in or out; this is usually performed while a white piece of paper is held slightly in front of the objective, and

(*b*) to focus the target or staff by means of the focusing screw which is a knob usually mounted on the barrel of the telescope. It is at this stage that parallax is tested for and, if necessary eliminated by delicate readjustment of the focusing screw.

Construction and setting up of levels

Dumpy level
The essential feature of this instrument, which is shown diagrammatically in Fig. 2, is that the telescope is constrained to move in a

plane at 90° to its axis of rotation. The direction of this axis is controlled by the levelling head, which comprises two so-called parallel plates separated by three or, in days gone by, four, footscrews. The lower parallel plate screws on to a tripod; the axis of rotation of the telescope is perpendicular to the upper parallel plate and may be made vertical, therefore, by correctly adjusting the footscrews. To set up the instrument ready for levelling, the bubble tube attached to the telescope is rotated until it is parallel to two of the footscrews (Fig. 3 (a)), and these are turned approximately equal amounts in opposite directions until the bubble is at the centre of its run. The bubble tube is then turned through 90° (Fig. 3 (b)) and the bubble centred by the third footscrew. These two operations are repeated, in turn, until the bubble is central in both positions (a) and (b).

If the bubble tube is in correct adjustment the telescope will now move in a horizontal plane, and the bubble will be central for every position of the telescope. That the bubble tube is not out of adjustment is easily tested by observing it both in position (a) and again at 180° from this position. The bubble should, of course, be central in both positions; if it is not, it is returned to centre (when at 180°) by making half the adjustment on the footscrews and half on the appropriate nuts, shown in Fig. 2, for adjusting the bubble tube.

It should be noted that "parallel plates" is something of a misnomer, for they will seldom be parallel when the instrument is set up.

Tilting level

Figure 4, showing part of a tilting level, indicates how, in contrast to the dumpy level, the telescope is not constrained to move parallel to the upper parallel plate.

The initial setting-up operation is simpler and quicker than for a dumpy level, and is as follows.

(a) Adjust the footscrews (some instruments have a ball-and-socket joint instead of footscrews) until the circular bubble, usually situated on the upper parallel plate, is central. The instrument is then roughly level.

(b) Point the telescope at the target, finally making the line of sight exactly horizontal by means of the tilting screw and the sensitive bubble tube mounted parallel to the telescope. In practice, the tilting screw is a highly geared differential screw permitting very fine movements of the telescope.

It is necessary to readjust the tilting screw for each different pointing of the telescope.

Precise level

Modern precise levels are usually tilting levels with certain additional

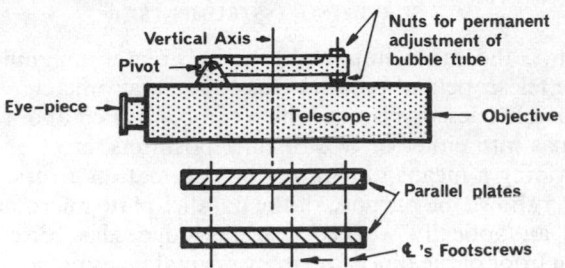

FIG. 2. *Essential features of a dumpy level.*

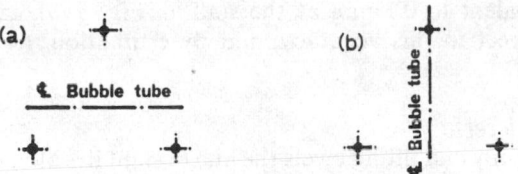

FIG. 3. *Setting up a dumpy level.*

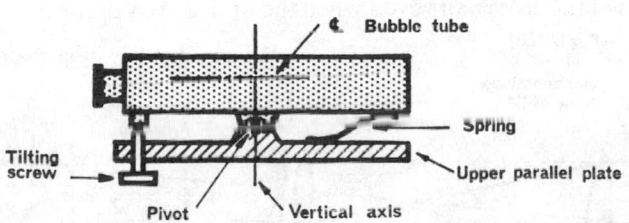

FIG. 4. *Essential features of a tilting level.*

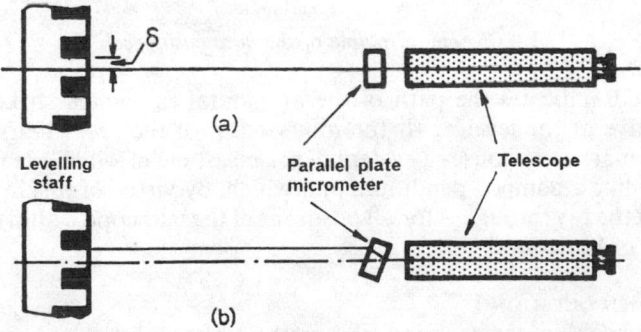

FIG. 5. *The operation of the parallel-plate micrometer.*

refinements, the more important being greater magnification, a reversible telescope and parallel glass plate micrometer.

A reversible telescope is one which can be rotated about its longitudinal axis into either of two distinct positions, erect or inverted, thus providing a means of eliminating collimation errors.

Figure 5 shows the purpose of the parallel-plate micrometer. This device is an optically worked, parallel-sided glass disc which is pivoted in front of the objective; in its normal position, Fig. 5 (a), the line of collimation is not affected. Rotation of the glass disc displaces the line of collimation parallel to itself by refraction, Fig. 5 (b). The device is operated by turning a drum which is calibrated in divisions, each equivalent to 0.1 mm at the staff. δ, Fig. 5 (a), can therefore be read direct to this accuracy, and by estimation, to the nearest 0.01 mm.

Automatic level

In both dumpy and tilting levels the line of sight lies along the optical axis of the telescope and is truly horizontal only when these instruments are correctly set up. The main feature of automatic levels is that they provide a horizontal line of sight passing through the intersection of the cross hairs even when the optical axis of the instrument is not horizontal.

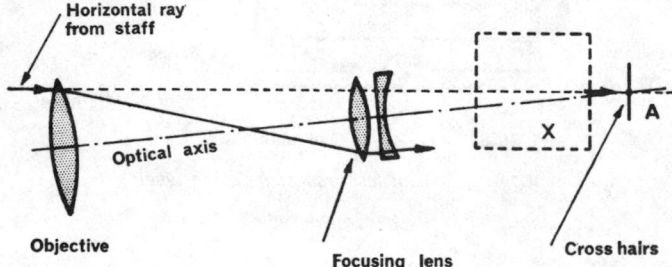

FIG. 6. *The principle of the automatic level.*

Figure 6 indicates the path of the horizontal ray which strikes the objective at the level of A, the intersection of the cross hairs. The space marked X houses a system of prisms, some of which are free to swing like a damped pendulum and which, by virtue of this motion, reflect the ray through A for all positions of the telescope within about $\pm 20'$ of horizontal.

SPECIMEN QUESTION 1

Distinguish, in construction and method of use, between "dumpy" and "tilting" levels.

State in general terms the principle of an "automatic" level.

SOLUTION
The difference in construction between dumpy and tilting levels is described on p. 4, and the principle of an automatic level is described above.

Methods of use of dumpy and tilting levels. The two types of instrument are set up for levelling by different procedures, as described above. In addition, they are suited to different types of survey. The dumpy level is the better instrument when a large number of staff readings are required from each instrument station, so that the longer time for setting up is compensated by the saving in time at each new telescope pointing. Examples are: (*a*) the setting out of a site prior to construction on it, and (*b*) contouring, in which the level may be used to obtain a large number of spot levels. Tilting levels are best suited to level runs in which only backsights and foresights (*see* Chapter 3) are taken from each instrument station, e.g. in transferring bench marks.

Construction and use of the vernier theodolite
This instrument will be described in two stages: (*a*) the parts used in measuring horizontal angles; and (*b*) the parts used in measuring vertical angles.

Horizontal angles
Figure 7 (*a*) shows, schematically, how the upper and lower clamps are related to the horizontal circle and verniers. In practice, a tangent

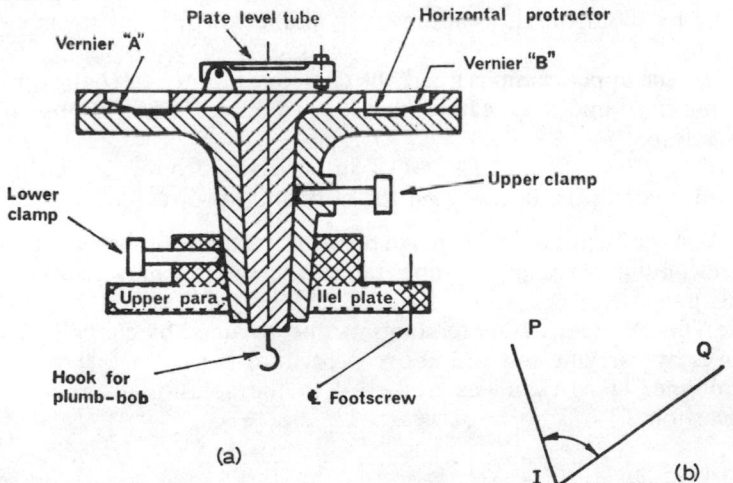

FIG. 7 (*a*) *Schematic representation of the parts of a theodolite used in measuring horizontal angles.* (*b*) *Measurement of horizontal angles.*

screw (not shown) for the fine control of horizontal rotation of the telescope is associated with each clamp, and can be used only when the clamp is tight.

A theodolite has two parallel plates separated by three or, in days gone by, four, footscrews just like a dumpy level. Setting up a theodolite is somewhat more difficult, however, for it involves two operations, either of which can to some extent upset the other. These are: (a) centring; and (b) levelling up.

(a) Centring involves positioning the instrument so that the plumb bob is exactly over the station, which may be marked by, for example, a small nail head. Usually a centring device, by means of which the vertical axis can be moved perhaps 5 mm (or more) in any direction, is incorporated either in the levelling head or at the head of the tripod to facilitate this operation.

(b) Levelling up. A theodolite is levelled in exactly the same way as a dumpy level. If the theodolite is to be used for measuring horizontal angles it is usually sufficiently accurate to level up by reference to the plate-level tube, which is less sensitive than the altitude-level tube described below.

The procedure for measuring horizontal angles is as follows. To measure $\angle PIQ$ in Fig. 7 (b):

(a) The theodolite is carefully set up at I, levelled and centred.

(b) The telescope is pointed to P, both clamps (see Fig. 7 (a)) are tightened, and horizontal scale readings are taken on verniers A and B (minutes and seconds only on B, since the degrees will in general differ by 180° from the reading on A) and meaned, giving a value of, say θ_1.

(c) The upper clamp is freed, the telescope pointed to Q, the upper clamp fixed and the mean reading θ_2 of verniers A and B observed as before.

(d) $\angle PIQ = \theta_2 - \theta_1$ (+360° if, in traversing from P to Q, the A vernier passes the 0° mark on the horizontal protractor).

At stage (b) fine adjustments in pointing are made on either tangent screw, but at stage (c) the upper tangent screw only may be used for this purpose.

[*Note.* More accurate readings can be obtained by changing face —i.e. by carrying out the above procedure twice: once "face left" and once "face right" as described in the solution to Specimen Question 2.]

Vertical angles
Figure 8 (a) shows the telescope and vertical scale mounted on the same shaft, the horizontal traverse axis or transit axis. This shaft can

turn in bearings mounted at the top of two brackets named, because of their shape, A-frames. Shaft, telescope and vertical circle always turn together.

Figure 8 (*b*) shows the T-frame. The horizontal axis passes through this frame. At each extremity the T-frame carries a vernier, V, against which the vertical scale is read. The T-frame is held in position by the clip-screws, which turn in projections from the A-frames; lock nuts (not shown) prevent the operator from accidentally altering the setting of the clip-screws.

Figure 8 (*c*) shows the means by which the telescope can be locked at any angle in the vertical plane. The horizontal axis passes through a boss; when the clamp is free the shaft can turn while the boss remains still. When the clamp is tightened down this movement is prevented, but the tangent screw then permits fine adjustment of the vertical angle. The tangent screw turns in a projection from the A-frame, and bears on a vertical limb, which is therefore rotated when the tangent screw is turned.

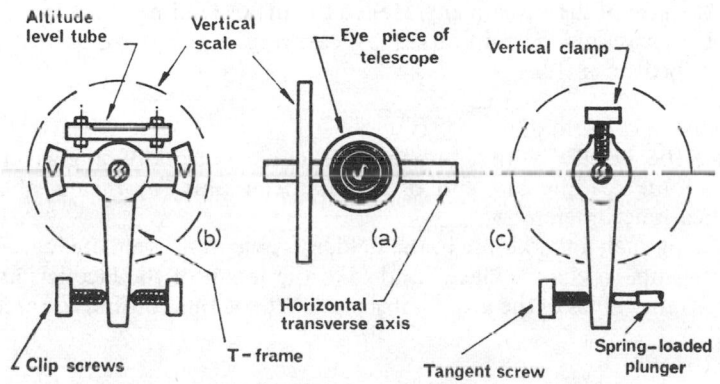

FIG. 8. *Schematic representation of the parts of a theodolite used in measuring vertical angles.*

The tangent screw turns in a projection from the A-frame, and bears on a vertical limb, which is therefore rotated when the tangent screw is turned.

The vertical circle level (labelled altitude level) is mounted on the T-frame and can be adjusted, relative to the T-frame, by means of a capstan nut and screw at one end of the tube, there being a pivot at the other end. The instrument is in adjustment only if the following three conditions can be satisfied simultaneously: the bubble is central: the vertical scale reads $0° \ 00' \ 00''$ and $180° \ 00' \ 00''$ against the verniers; the line of sight is horizontal.

SPECIMEN QUESTION 2

Describe, with diagrams, the relationship on a theodolite between the telescope, vertical circle, vertical circle level and their respective clamps and adjusting screws.

If, in measuring a vertical angle, a consistent difference of 02′ were recorded between the face-left and face-right readings, would the mean value be acceptable? How would you eliminate or minimise the difference?

SOLUTION

The text immediately above answers the first part of the question.

Figure 8 (*a*) shows the telescope face left, i.e. with the vertical scale to the left of the telescope when the observer is looking through the eyepiece. If the telescope is transitted through 180°, i.e. turned in a vertical plane, it is then inverted and, from the eyepiece, the vertical scale appears on the right, so the instrument is face right.

A consistent error of 02′ between face-left and face-right readings suggests a collimation error, which will be "up" or "down" depending on the face of the instrument. Hence the mean reading is acceptable. The error could be minimised by carrying out a two-peg test as described on p. 13.

SPECIMEN QUESTION 3

Describe briefly, with sketches, the construction of a glass-arc theodolite reading one end of the diameter only, by means of an optical micrometer.

Using such a theodolite in the field, it is essential, even after careful setting up, to change "face" and take the mean of the face-left and face-right values of the angles observed. Why is this routine adopted?

SOLUTION

Except for the means of reading the horizontal and vertical scales, the glass-arc theodolite is similar in construction to the vernier theodolite: its telescope can turn about the horizontal axis for the measurement of vertical angles, and about the vertical axis for the measurement of horizontal angles. It has three clamp and tangent screw arrangements, a levelling head and a centring device.

The principle of the optical micrometer is shown in Fig. 9. Figure 9 (*a*) shows light from a bright source (bright sky or artificial light) being reflected from an adjustable mirror, *A*, through an optically worked, parallel-sided, rotatable glass disc, *B*, thence through the glass-arc, *C*, to the observer's eye, *D*. In practice, the path *BCD* is reflected by prisms through a number of right-angles to keep it within the metal structure of the instrument.

Figure 9 (*b*) is an enlarged view of *B* and *C*. It indicates how, by

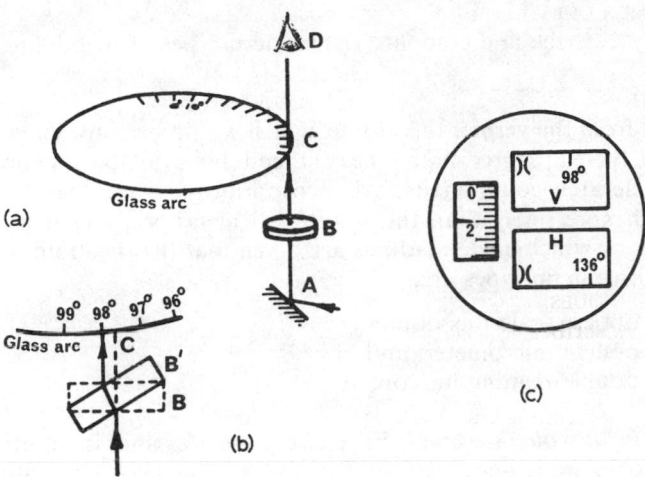

FIG. 9. *The principle of the optical micrometer.*

rotating *B* to *B'*, the incoming ray is deflected by refraction on to a whole degree mark—98° being illustrated. The angle through which *B* is rotated is a measure of the minutes and seconds, i.e. the amount by which the line of the ray of light differs from 98". The position of *B* is set by a knob called the micrometer control.

Figures 9 (*a*) and (b) can equally well be horizontal or vertical angle systems. In this type of theodolite the two optical systems are usually brought to the auxiliary eyepiece along a common path, so that vertical and horizontal angles can be read without moving the eye. Figure 9 (*c*) is an example of what might be seen in the auxiliary telescope. To read the vertical angle, *V*, the micrometer control is turned until the 98° mark is exactly between the "brackets"; as the micrometer control also moves the scale in the left-hand window, this scale gives the minutes and seconds, and therefore the complete vertical angle. Similarly, by further manipulation of the micrometer control, bringing the 136° mark between the brackets on the H-scale, the horizontal angle is read.

Changing face. This is necessary to eliminate errors due to:

(*a*) Faulty centring of the protractors;

(*b*) The line of sight not being perpendicular to the horizontal axis;

(*c*) The horizontal transverse axis not being perpendicular to the vertical axis; and

(*d*) A vertical circle index error.

SPECIMEN QUESTION 4

Briefly describe and compare the modern types of theodolite.

SOLUTION

Apart from the vernier theodolite, which would be considered traditional, there are three main types of theodolite available for common, or moderately common, use. All incorporate glass arcs, which can be etched more finely than the traditional metal arc, and it is in the means by which angle readings are taken that the theodolites differ. The three main types are:

(a) optical scale theodolites;
(b) optical micrometer; and
(c) double-reading micrometer.

(a) *Optical scale theodolites* have two glass arcs and, in addition, at least one, sometimes two, minor scales. The glass arcs are subdivided into intervals of, usually, 1° the range of the minor scale being equal to 1 division of the main scale. The minor scale is divided into subdivisions of 20 or 30 seconds, or sometimes 1 minute. Inside the theodolite an optical system ensures that the minor scale reading can be read against one of the marks on the main scale as indicated schematically in Fig. 10.

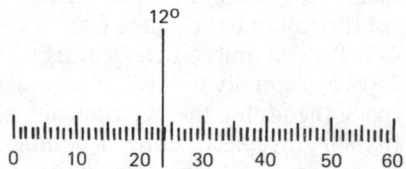

FIG. 10. *Reading an optical scale.*

Both vertical and horizontal scales are viewed through the same auxiliary telescope. In some instruments the two halves of the horizontal scale appear separately in the auxiliary telescope so that the reading recorded is equivalent to the mean of the two vernier readings on a vernier theodolite.

These instruments are the most straightforward in regard to the ease with which scale readings can be taken and the readings can be estimated fairly confidently to the nearest half divison of the minor scale—i.e. to the nearest 15 seconds or so.

(b) *Optical micrometer theodolites* are described fully in Specimen Question 3. Angle readings can be estimated to the nearest 5 seconds or so.

(c) *The double-reading micrometer theodolite* is a more refined version of the optical micrometer theodolite. The internal optical system is more elaborate and averaged through the optical micrometer to eliminate the possibility of index error. The better instruments of this type are capable of direct readings to the nearest 1 second with the possibility of estimating to the nearest half-second. Instruments in this class of accuracy are sometimes described as "universal" theodolites. Broadly speaking, the more accurate the theodolite the larger are the glass arcs and the greater the magnification of the telescope.

A variety of other refinements to theodolites are, also, now available. One particularly worth mentioning is the automatic vertical circle index mark which, in a sense, is the theodolite equivalent of the automatic level. This device compensates for non-verticality of the theodolite during use and, in fact, it is sufficient for it to be set up within about 3° of the true vertical. Setting up is, therefore, quicker and less demanding and is done with reference to a circular bubble. One method of constructing the automatic vertical circle index mark is by means of a damped pendulum arrangement carrying at its lower end a prism through which the vertical scale is viewed; the pendulum compensates for non-verticality and the movement of the prism eliminates the index error.

Permanent adjustments

To the dumpy level

In order that a dumpy level functions properly it is necessary that the axis of rotation of the telescope is vertical when the bubble is central, and that the line of collimation is perpendicular to this axis. The following tests will reveal whether any correction is necessary; if so, the adjustments are described as "permanent", and, although they are not permanent, they should not need repeating very often.

(a) *Adjustment of bubble tube.* A suitable procedure is described on p. 4.

(b) *Adjustment of line of sight*—the two-peg test. The procedure is as follows.

(i) The instrument is set up exactly midway between two pegs A and B, say 50 m apart, and readings are taken on a staff held at each in turn (*see* Fig. 11 (a)). Any error in the line of sight is angular. Therefore, since A and B are equidistant from the telescope, the error in the staff reading, e, will be the same at each station.

Hence, if a_1 and b_1 are the staff readings observed the true difference in level between A and B is

$$(b_1 + e) - (a_1 + e) = b_1 - a_1$$

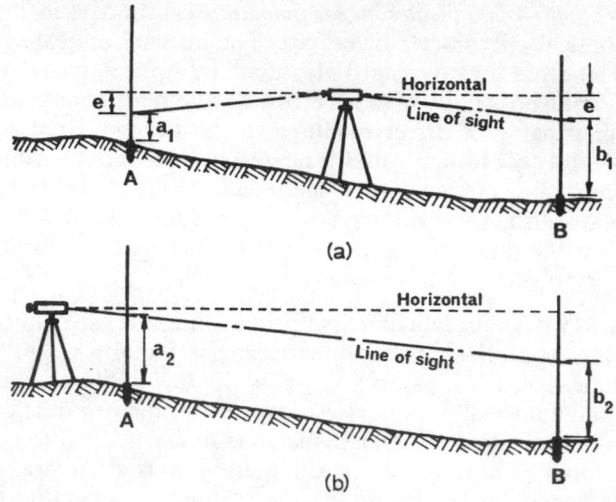

FIG. 11. *The two-peg test.*

(*ii*) The instrument is then moved to a new position not more than about 3 m from A, but far enough from A to be able to focus a staff held there, Fig. 11 (*b*). Staff readings a_2 and b_2, at A and B, are observed.

If $b_1 - a_2 \neq b_1 - a_1$ the diaphragm must be raised or lowered until the line of sight makes a reading at B of $a_2 + b_1 - a_1$. Since the error at A, because of the proximity of the instrument and staff, is usually negligible, the collimation should now be perpendicular to the vertical axis. It is essential, however, always to retest the instrument after making an adjustment, for the delicate setting of the instrument can be disturbed by the slightest clumsiness when making the adjustment—in which case it might easily be over-corrected, or under-corrected, to quite a marked extent.

For a tilting level only one test is necessary, for it is sufficient that the line of sight is horizontal when the bubble is central. Details of this test (again a two-peg test) are left as an exercise for the reader.

To the vernier theodolite
Usually, four tests are carried out, their order being important. They are as follows:

(*a*) *Plate bubble tubes.* Most theodolites have two plate bubble tubes. The more sensitive should be tested and adjusted in the same way as a dumpy-level bubble tube, the other being adjusted from it.
(*b*) *Horizontal transverse axis* (or trunnion axis or transit axis). This

axis must be perpendicular to the vertical axis. The simplest test is to employ a striding level, if available, to detect if the trunnion axis is not horizontal when the instrument is correctly levelled for use. This device comprises a sensitive level tube mounted on two machined legs designed to "stride" the telescope and vertical circle, and to stand on the horizontal axis. When it is employed the striding level itself should be tested by reversing it end for end.

(c) *Line of sight.* Both horizontal and vertical cross hairs should coincide with diameters of the objective. A simple method, though not the best, of testing the horizontal cross hair (less important than the vertical cross hair) is to set up with the telescope horizontal, observe a levelling staff and adjust (if necessary) the cross hair so that it gives the mean reading of those at the top and bottom of the field of view.

Referring to Fig. 12, the vertical cross hair is tested by the following procedure.

(*i*) The instrument is set up at *I* on a level field, and the telescope sighted on to a fine mark at *A*, about 30 m away, both horizontal clamps being tightened.

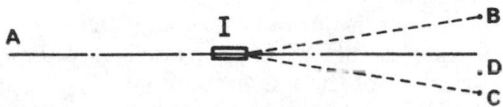

FIG. 12. *Adjusting the vertical cross hair.*

(*ii*) The telescope is transitted and a fine mark made at *B*, also perhaps 30 m from *I*, where the line of sight falls.

(*iii*) By freeing either horizontal clamp, the telescope is traversed to resight *A*, after which both horizontal clamps are fixed.

(*iv*) The telescope is again transitted and a second mark made at *C*, on the line of sight, alongside *B*.

If *BC* is appreciable the diaphragm must be adjusted so that the line of sight is moved from *C* to *D*, where $CD = \frac{1}{4}CB$. It should be noted that the distances *AI* and *IB* need not be equal.

(d) *Vertical circle index error, and altitude bubble tube.* A two-peg test is carried out, the instrument being levelled in each position by reference to the altitude bubble, which is more sensitive than the plate bubble. All staff readings are taken with the same vertical scale reading, namely the reading which should correspond to a horizontal telescope ($0°$ on some instruments, $90°$ on others). When the instrument is set up as in Fig. 11 (*b*) the required distant staff reading can be calculated as for the dumpy level and compared with the actual distant staff reading. If these differ, adjustments are made in the following order.

(*i*) Adjust the line of sight by the vertical tangent screw to give the calculated staff reading. This will upset the vertical scale reading.

(*ii*) Restore the vertical scale reading to 0° (or 90°) by moving the verniers by means of the clip screws (*see* Fig. 8 (*b*)). This will upset the altitude bubble.

(*iii*) Centralise the altitude bubble by means of the adjusting nuts at one end.

Remember, after every test and adjustment, the instrument should be re-tested.

EXAMINATION QUESTIONS

1. State which errors of adjustment in a theodolite are eliminated by taking the mean of face-right and face-left readings. Assume all readings are the mean of both verniers or both micrometers.

Describe how you would test for, and adjust, any two of these errors.

2. In measuring a horizontal angle with a theodolite between two directions, one lying in the horizontal plane and one at an elevation of say 60°, errors may arise from lack of adjustment of collimation axis, transit axis and vertical axis. Explain briefly how these errors will affect the measurement of the horizontal angle.

How can such errors be detected and what procedure will eliminate any or all of their effects?

3. List the four chief "permanent" adjustments to a theodolite that you would make in the order in which they should be made.

Describe in detail how you would test for any vertical circle index error and describe how you would make the adjustment in the following cases:

(*a*) Vernier clip and vertical circle tangent screw on separate arms of the theodolite;

(*b*) Vernier clip and vertical circle clamp and tangent screw on the vernier frame.

Is the index error eliminated by taking the mean of face left and face right?

[*Note.* Theodolites constructed as in (*b*) are rare nowadays. The procedure for the test on these instruments differs only slightly from that described on p. 15, can be found in, say, Wilson's *Land Surveying* if the reader is unable to think it out for himself.]

4. With the aid of diagrams, show how an angle of elevation above the horizontal is measured on a theodolite.

How would you check that such angles are related to a true horizontal datum?

What is the difference in the lengths of the radii of curvature of the vertical circle bubbles of two theodolites in which the equivalent of one division on the level tube, 2.5 mm long in each case, are 18″ and 30″ respectively?

[*Hint:* In the first case find the radius of the circle for which a chord of length 2.5 mm subtends an angle at the centre of 18″.]

5. Two stations, A and B, are 30 m apart. A level was set up 3 m away from A on the opposite side from B, and the readings taken on a levelling staff held at A and B were 2.560 m and 2.080 m. The instrument was then moved to a station midway between A and B and the readings taken on to a staff held at A and B were 1.900 and 1.480 m respectively.

Was the instrument in adjustment? If not, by how much was it out of adjustment and how would you correct it?

6. State the basic difference in the design of the dumpy and the tilting level. Comment on the effect of this difference on the permanent adjustments of the instruments and describe the test for the permanent adjustment common to both levels.

7. The difference in level between point A at the top and point B at the bottom of a uniform slope was found to be 12.25 m. The slope was 240 m long and the dumpy level was used so that the foresights were twice as long as the backsights.

A "two peg" test was later carried out on the instrument giving the following staff readings: 2.25 m on peg X and 2.39 m on peg Y with the instrument half-way between the pegs, and 2.31 m and 2.48 m on pegs X and Y, respectively, with the instrument close to peg Y.

Determine the actual difference in level between points A and B, given that distance $XY = 60$ m.

8. Explain how a dumpy level may be set up with its axis of rotation vertical, even if the bubble tube is not in perfect adjustment.

A dumpy level in this condition is set up, for test purposes, with its axis vertical, midway between two staves A and B which are 80.000 m apart. The reading on staff A is 0.616 m and that on staff B 1.964 m. The level is then set up adjacent to staff A, again with its axis vertical. The reading on staff A, by reverse sighting through the objective, is 1.601 m and the reading on B 2.748 m.

If the bubble is mounted on the telescope tube and adjusting screws are provided for the diaphragm and the bubble tube, describe the completion of the test and adjustment of the instrument, inserting staff readings where appropriate.

CHAPTER 2

BASE-LINE MEASUREMENT

SETTING OUT

Before developing a new site, for example the site of a proposed steel works, it is necessary to establish a works base-line by placing a number of pegs in a straight line. All future construction, be it building roads, sewerage, etc., can then be correctly positioned and correlated by making suitable setting-out measurements either directly or indirectly from the base-line. The setting-out pegs can be placed by:

    (*a*) measuring lengths only;
    (*b*) measuring angles only (i.e. by triangulation);
    (*c*) measuring angles and lengths;
    (*d*) chaining and offsetting.

Whichever method is used, however, the accuracy of the completed work cannot exceed the accuracy of the base-line measurement, which should therefore be the most accurate measurement in the construction programme.

SURVEYING

A map or plan of an existing area can be prepared only after taking a great number of site measurements from which it is possible to plot the position of all the important features. Again, the recorded observations may be:

    (*a*) linear measurements (i.e. chain surveying—for small areas);
    (*b*) angular measurements (i.e. triangulation);
    (*c*) linear and angular measurements (i.e. traversing).

In chain surveying and traversing, lengths are usually measured by chain or by a method of comparable accuracy. Triangulation, involving one linear measurement, the base-line, is used when the highest degree of accuracy is required, but it follows that the accuracy of the survey cannot exceed the accuracy of the base-line measurement.

USE OF THE STEEL BAND

From the foregoing paragraphs it is apparent that base-line measurement entails a high degree of accuracy, the permissible error depending, of course, on the particular job in hand. For the best work, an invar steel (expensive) band or wire is used, invar steel having the lowest coefficient of expansion of all known metals. Frequently, however, mild steel bands or wires (or even tapes) are used in engineering surveys.

Standardisation

A steel band is an expensive piece of equipment which will have been carefully standardised before being sold. It bears two end marks nominally 30 m (or other suitable distance) apart. The band is said to be standardised when the length between the end marks has been accurately determined under carefully controlled conditions of temperature and tension. For this purpose it is usually laid flat in a horizontal test bay. By contrast, field measurements are usually made as described in Chapter 1 with the band hanging freely between two tripods, which probably are not at mean sea level (msl). On site there may, too, be a difference between the reduced level at each end of the band, and the average temperature and pull may differ from the temperature and pull during standardisation.

TABLE 1. STANDARD AND FIELD CONDITIONS

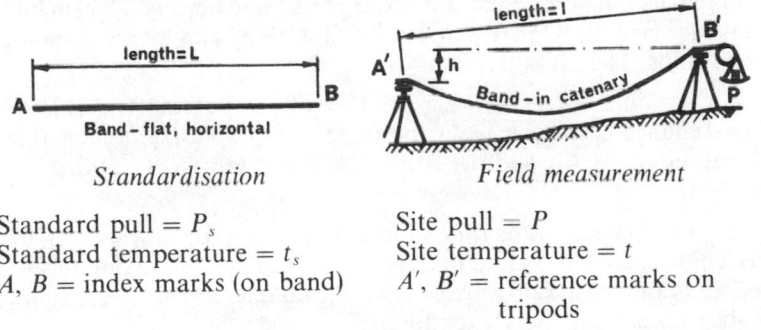

Standardisation	Field measurement
Standard pull $= P_s$	Site pull $= P$
Standard temperature $= t_s$	Site temperature $= t$
A, B = index marks (on band)	A', B' = reference marks on tripods

Base-line measurement amounts, therefore, to a comparison of measurements made under quite different conditions, which are summarised in Table 1. The difference between standard and field conditions will be as follows.

(a) A and A' can be made to coincide in plan, but then B and B' will not usually coincide.

(b) In general L and l will differ, but only by a little.

(c) If P is increased the band will "stretch".

(d) If t is increased the band will expand, becoming longer.

In order to determine the true value of the horizontal component of l, reduced to msl, the following corrections must be applied to the observed length, $A'B'$ measured along the band.

Corrections for field conditions

Notes on the proofs of the following formulae are given on pp. 25–6. Let:

W = weight of band between supports (N).

A = cross-sectional area of the band (m^2).

E = Young's modulus of elasticity of the band (N/m^2).

α = coefficient of linear expansion.

R = radius of the Earth.

H = mean height of the supports above msl (in the same units as R).

Then:

(a) *sag correction* = $L/24.\ (W/P)^2$ (subtract)

(b) *pull correction* = $(P - P_s)L/(AE)$ (add)

(c) *temperature correction* = $(t - t_s)\alpha L$ (add)

(d) *slope correction* = $h^2/(2L)$ (subtract)

(e) *altitude correction* = HL/R (subtract)

In these correction formulae it is convenient to redefine L as the nominal length of the band. For example, a band nominally 30 m long may measure 30.042 m (*see* Specimen Question 7) under the standard conditions. Then it is necessary to apply:

(f) *standard correction*, e.g., in this case, 0.042 m. The rule for the standard correction is: if the band is too long (i.e. longer than the nominal length), add the correction; if it is too short, subtract the correction.

A base-line may be as much as ten kilometres (or more) in length. Its length is then measured in bays, each bay being approximately the length of the band. Obviously all bays should lie on one straight line; if they do not it is necessary to apply:

(g) *alignment correction*, which can be calculated by the slope correction formula if h is taken as the amount by which the bays are out of line.

The directions "subtract" or "add" are applied as follows:

true length = observed length $- L/24.(W/P)^2 + (P - P_s)L/AE \ldots$

The correction formulae as given apply exactly, providing the standardising routine and fieldwork routine are as described in this chapter. It is fairly easy for an examiner to describe a slightly different procedure which necessitates modifying these formulae. The worked specimen questions illustrate this point.

A steel tape is 30 m long between end graduations at a temperature of 20° C and under a pull of 44.5 N when lying on the flat. Its cross-sectional area is 6.5 mm^2, its mass is 1.5 kg and its coefficient of linear expansion 0.000011 per °C.

The tape is stretched over two supports between which it records 30.000 m and is supported at two intermediate supports equally spaced, all supports being on the same level and the tape being allowed to sag freely between supports.

The temperature at observation is 25° C and the pull on the tape is 70 N. Calculate the actual length between the end graduations, and its equivalent length at msl assuming the measurement referred to was made at 2000 m.

$$E = 2 \times 10^{11} \text{ N/m}^2 \qquad R = 6\,370\,000 \text{ m}$$

SOLUTION
Since mass = 1.5 kg, Weight, $W = 1.5 \times 9.81 = 14.7$ N

As there are three (equal) bays, three (equal) sag corrections must be applied, one for each bay. For this purpose, therefore, L must be taken as 30/3, W as 14.7/3, but P must be taken as 70 N, being the same in each bay. To calculate the temperature and pull corrections, however, notice that L is taken as 30 m, the nominal length of the tape.

	Error		*Correction*
Sag	$= 3 \times \dfrac{30}{3 \times 24}\left(\dfrac{14.7/3}{70}\right)^2$	$=$	$-0.006\,12$
Temp.	$= (25 - 20)\,30 \times 0.000\,011$	$= $	$+0.001\,65$
Tension	$= \dfrac{(70 - 44.5) \times 30}{2 \times 10^{11} \times 6.5 \times 10^{-6}}$	$=$	$+0.000\,59$
		Totals	$+0.002\,24 \quad -0.006\,12$

hence Actual length $= 30.000 + 0.002\,24 - 0.006\,12$
$= 29.996$ m (to three decimal places)

Correction for altitude $= 2000 \times 30/6\,370\,000 = 0.009\,43$ m
$= 0.009$ to three decimal places
∴ Corrected length at msl $= 29.996 - 0.009$
$= 29.987$ m

In the next example, notice that the procedure described in the text has been reversed in so far as the band is standardised in a catenary but on the site is used flat. Consequently, the sag correction is added.

Another point to observe is that the pull correction is negative, P_s being greater than P, so that a negative quantity is "added".

SPECIMEN QUESTION 6

Two brass plates with finely marked lines have been set into the floor of a building and have been used as a "30-m standard" for some time. In remeasuring this distance a steel tape was laid flat on the floor, and with a pull of 65 N and a temperature of 22° C the recorded distance between the marks was 29.942 m. The tape had previously been standardised in catenary and measured 30 m at 70° C under a pull of 105 N. What was the true length between the marks on the floor to the nearest 0.001 m? Weight of tape = 15 N; sectional area = 3 mm²; coefficient of expansion = 0.000 012 per °C; $E = 200$ GN/m².

SOLUTION

	Error	*Correction*	
Sag	$= \dfrac{30}{24} \times \left(\dfrac{15}{105}\right)^2$	$= +0.0255$	
Pull	$= \dfrac{(65 - 105) \times 30}{3 \times 10^{-6} \times 2 \times 10^{11}}$	$=$	$-0.002\,00$
Temp.	$= (22 - 7) \times 0.000\,012 \times 30$	$= +0.0054$	
	Totals	0.0309	$-0.002\,00$
		-0.0020	
		0.0289	
Add observed length		29.942	
True length (to nearest mm)		$= \underline{\underline{29.971\,\text{m}}}$	

NOTES

Where it can be performed quickly, long division and multiplication giving an exact answer is the best method of calculating the corrections, of course. This method is not convenient for the first correction, however, and consequently a slide-rule was used to calculate the sag correction. In general a slide-rule may be expected to give only a small error in the third significant figure (i.e. the fourth decimal place in the sag correction) and 4-figure logs only a small error in the fourth significant figure. The methods of calculation should be of comparable accuracy and the choice between the slide-rule and logs must therefore be made with some care. The required accuracies described above also apply if a calculator is used.

The same principles apply in all such base-line calculations.

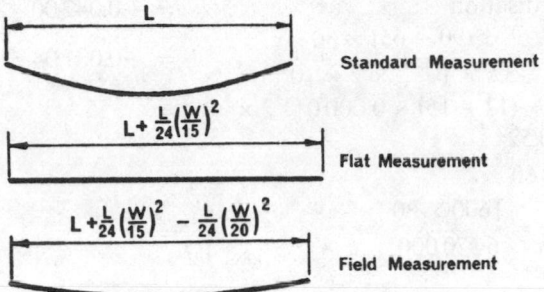

FIG. 13. *Sag correction from standard catenary to site catenary.*

$1 \text{ GN/m}^2 = 10^9 \text{N/m}^2$

The complication of reversing the sign of the sag correction in the previous specimen question is taken a stage further in the next, in which the band is standardised in a catenary under a pull of 90 N. Fig. 13 indicates the technique adopted.

It is necessary to apply two corrections, one to be added to, and one to be subtracted from, the observed length.

SPECIMEN QUESTION 7

The details given below refer to the measurement of the first "30-m" bay of a base-line. Determine the correct length of the bay reduced to mean sea level.

With the tape hanging in catenary at a tension of 90 N and at a mean temperature of 12° C, the recorded length was 30.0824 m. The difference in height between the ends was 0.52 m and the site was 1600 m above msl.

The tape had previously been standardised in catenary at a tension of 65 N and at a temperature of 15° C, and the distance between zeros was 30.042 m.

$R = 6\,370\,000$ m. Mass of tape $= 0.024$ kg/m. Sectional area of tape $= 3.6$ mm^2. $E = 2 \times 10^{11}$ N/m^2. Temperature coefficient of expansion of tape $= 0.000\,011\,2$ per °C.

SOLUTION

Since a mass of 1 kg has a weight of 9.81 N,
Weight of tape $= 30 \times 0.024 \times 9.81 = 7.06$ N

Error		*Correction*

$$\text{Sag (1)} = \frac{30}{24} \times \left(\frac{7.06}{65}\right)^2 \qquad\qquad = +0.014\,75$$

$$\text{Sag (2)} = \frac{30}{24} \times \left(\frac{7.06}{90}\right)^2 \qquad\qquad\qquad\qquad\quad -0.007\,69$$

$$\text{Standardisation} \qquad\qquad\qquad\qquad = +0.042\,00$$

$$\text{Pull} \quad = \frac{(90-65) \times 30}{3.6 \times 10^{-6} \times 2 \times 10^{11}} \quad = +0.001\,04$$

$$\text{Temp.} = (12-15) \times 0.000\,011\,2 \times 30 \quad = \qquad\qquad -0.001\,01$$

$$\text{Slope} \quad \frac{0.52^2}{60} \qquad\qquad\qquad\qquad\qquad = \qquad\qquad -0.004\,51$$

$$\text{Altitude} = \frac{1600 \times 30}{6\,370\,000} \qquad\qquad = \qquad\qquad -0.007\,54$$

	Totals	0.057 79	−0.020 75
		−0.020 75	
		0.037 04	
Add observed length		30.0824	
True length (to four dec. places)		30.1194 m	

NOTES

In the slope correction, $0.52^2 = 0.2704$ exactly. This can be obtained from tables or calculator or, alternatively, can be fairly rapidly calculated thus:

$$0.52^2 = (0.5 + 0.2)^2 = 0.5^2 + 2 \times 0.5 \times 0.02 + 0.02^2$$
$$= 0.25 + 0.02 + 0.0004$$

In Specimen Questions 6 and 7, L is taken as the nominal band length, namely 30 m, in every correction formula. A second-order error results, for it would be more accurate to use the observed length for L. For example, in Specimen Question 7 it would be more accurate to take the slope correction as $0.52^2/(2 \times 30.0824)$. But this differs from $0.52^2/60$ by approximately 0.27%, since 30.0824 is approximately 0.27% greater than 30.

$$\text{Since} \qquad\qquad 0.52^2/60 = 0.004\,51,$$
$$0.52^2/2 \times 30.0824 = 0.004\,51 - 0.27\% \times 0.004\,51$$
$$= 0.004\,51 - 0.000\,012$$

0.000 012 is the second-order error and is negligible, as it is in most examples. There are occasions, of course, when the second-order error must be taken into account—for example, when a 30-m tape is used to measure a "short" bay of, say, 29.5 m.

Derivation of the correction formulae

(a) *Sag correction.* The formula $L/24 . (W/P)^2$ is derived at some length from the geometry of either the catenary or the parabola. Both proofs are given in Wilson's *Land Surveying*, for example, but not here, as it is considered they would obscure the main theme of this chapter.

(b) *Pull correction.* The formula $(P - P_s)L/(AE)$ follows immediately from the definition of Young's modulus:

$$E = \frac{\text{Stress}}{\text{Strain}} = \frac{(P - P_s)}{A} \cdot \frac{L}{\text{Correction}}$$

(c) *Temperature correction.* The formula $(t - t_s)\alpha L$ follows immediately from the definition of the coefficient of linear expansion, namely, α = fractional increase in length per degree rise in temperature.

(d) *Slope correction.* The formula $h^2/(2L)$ is obtained by considering Fig. 14, in which L is the length measured along the slope, and L' is the corresponding plan dimension, i.e. the dimension which would be used in preparing a plan or map.

$$\begin{aligned}
\text{Correction} = L - L' &= L - \sqrt{L^2 - h^2} \\
&= L - L(1 - h^2/L^2)^{\frac{1}{2}} \\
&= L - L(1 - \tfrac{1}{2}h^2/L^2 \ldots)
\end{aligned}$$

by the binomial theorem. Hence, to a first approximation,

$$\text{Correction} = h^2/(2L)$$

Notice that the omission of the next term $1/8 . h^4/L^3$ constitutes another type of second-order error. It is left as an exercise for the reader to show that if this correction is made in Specimen Question 7 it leads to a second-order correction of 0.000 000 34 m.

If the slope is given as an angle, θ radians, *see* Question 1 on p. 33.

$$\begin{aligned}
\text{slope correction} &= L(1 - \cos \theta) \\
&= L - L\left(1 - \frac{\theta^2}{2} + \ldots\right) \\
&= \frac{1}{2}L\theta^2 \text{ approximately.}
\end{aligned}$$

(e) *Altitude correction.* Figure 15 shows the length L measured at a

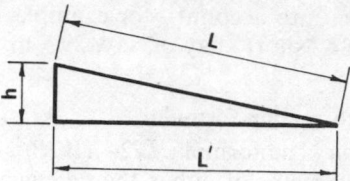

FIG. 14. *Slope correction.*

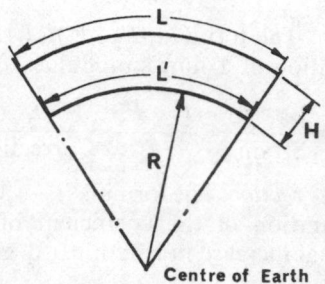

Centre of Earth

FIG. 15. *Altitude correction.*

height H above msl and the corresponding measurement L' at msl.
For the purpose of plotting maps or plans L' is the required
dimension, so that the correction is $L - L'$

but $$L/L' = (R + H)/R = 1 + H/R$$

i.e. $$L = L' + HL'/R$$

and $$L - L' = HL'/R = HL/R, \text{ approximately}$$

(*f*) *Standard correction.* Lastly, the standard correction rule is
easily verified by cutting a straight stick to a length of, say, 0.310 m,
calling it a nominal 0.300 m, using it to measure a distance between
two points marked on a flat bench, and comparing the result with
that obtained by making the same measurement with a standard rule.

ELECTROMAGNETIC DISTANCE MEASUREMENT

With the advance made in electronics and technology in general it
has become possible since about 1950 to measure distances by observing certain features exhibited by electromagnetic waves which have
been propagated under carefully controlled conditions. The technique is known as electromagnetic distance measurement, or E.D.M.
as it is often called, and it is now replacing the traditional methods
of base-line measurement using an invar band. At its best the accuracy

obtainable is of the same order as for angle measurements in geodetic surveys. The production of E.D.M. instruments is still a developing industry with a continuing quest for lighter and simpler apparatus and, because of this, it is possible here only to describe the more significant developments which have already occurred and give an indication of the principles underlying E.D.M.

The *geodimeter* was the first instrument to appear and has been continually improved since. In the earlier models, for example, the carrier wave was a light beam from a conventional tungsten filament whereas in the latest models a laser beam is used. The geodimeter is used in conjunction with a reflector at the distant station, the light ray returning to the geodimeter where the readings are taken.

The *tellurometer* was developed after the geodimeter and has been described as the most important development in surveying since the introduction of invar. It uses radio waves as the carrier. Two units are used, one each end of the base-line. One, usually called the *master*, transmits a signal which is processed at the distant instrument, which is usually described as *remote*, and returned to the master where the readings are taken. Improvements have been made continuously since the first model and now several models exist which are capable of operating over various distance ranges from about 3 km up to 100 km or more.

Other well-known instruments are the Wild distomat which uses an infra-red carrier wave and the mekometer which uses polarised light and which has a very high accuracy over its relatively limited range of 3 km.

The instruments described above use a variety of different carrier waves, all of which are electromagnetic. All share the basic principle that distance is measured by making observations on the carrier wave when it leaves the transmitter and again when it returns to the transmitter. In effect, the time taken for it to make the double journey is measured in the modern equivalent of the way in which Fizeau measured the velocity of light in 1849. To accomplish this the wave has to be given features which can be recognised both at the beginning and end of its journey and this is achieved by superimposing on it another signal—a process known as modulation. The fundamental differences between the various E.D.M. instruments lie (*a*) in the nature of the carrier wave used, (*b*) the practicalities of modulating it and (*c*) the practicalities of making the measurements.

Basic principles

All electromagnetic waves travel with the speed of light, v, which is very nearly $300\,000$ km/s, the approximate value used throughout the rest of this text. v varies slightly with variations in air temperature and barometric pressure.

All electromagnetic waves are sinusoidal, the usual way of depicting the generation of a sine wave being shown in Fig. 16.

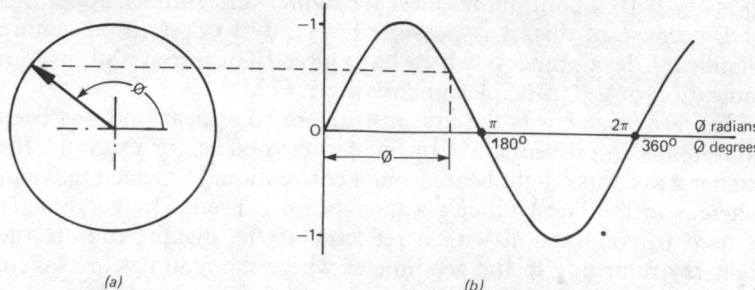

FIG. 16. *Generation of a sine wave. The radius vector in (a) rotates counter-clockwise at a uniform rate. If its vertical projection is plotted against ϕ, as in (b), a sine wave is produced.*

$\emptyset$ = phase angle.   In 1 revolution $\emptyset$ goes from 0 to 2π radians and, although it could be thought of as increasing by 2π with each revolution, the multiples of 2π are ignored and $\emptyset$ is expressed, at any stage, as an angle between 0 and 2π.

The number of revolutions per second at which the radius vector in Fig. 16(a) rotates is called the frequency, f, and this is related to the wavelength, λ, by the equation:

$$\text{velocity of light, } v = f \times \lambda$$

Modulation

The concept of modulation can be illustrated by referring to two tuning forks which vibrate with different frequencies $f_1 = 500$ Hz, say, and $f_2 = 490$ Hz. If these are made to vibrate in the same plane the oscillations will sometimes be in phase, when the combined note will be loudest, and sometimes out of phase, i.e. moving in opposite directions, when the combined note will virtually disappear.

This produces a *beat* frequency of $f_1 - f_2 = 10$ Hz.

The beat frequency is a modulated wave.

Another example is given in Fig. 17 which shows in (a) a frequency of 4 Hz, in (b) a frequency of 3 Hz and in (c) the modulated wave with a frequency of $4 - 3 = 1$ Hz.

At $A - A$ in Fig. 17.

$$\emptyset_1 = 1.67\pi$$
$$\emptyset_2 = 0.75\pi$$
$$\therefore \emptyset = \emptyset_1 - \emptyset_2 = 0.92\pi$$

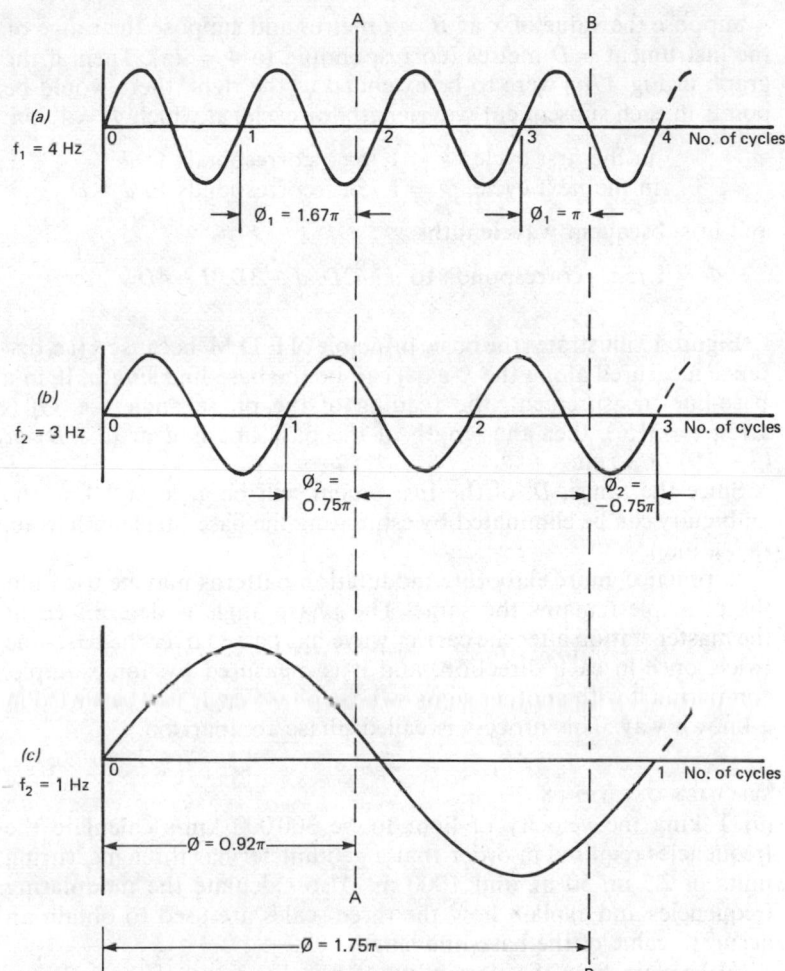

FIG. 17. *A simple modulated wave (c) resulting from the combination of the two other sine waves in (a) and (b).*

At $B - B$

$$\emptyset_1 = \pi$$
$$\emptyset_2 = -0.75\pi$$
$$\therefore \emptyset = \emptyset_1 - \emptyset_2 = 1.75\pi$$

The modulated frequency in Fig. 17 (c) is such that in the range shown, $\emptyset$ lies everywhere between 0 and 2π radians and, therefore, every value of $\emptyset$ corresponds uniquely to a value of the distance, x, measured along the x-axis.

Suppose the value of x at $B = d$ metres and suppose the range of the instrument $= D$ metres (corresponding to $\Phi = 2\pi$). Then, if the graph in Fig. 17(c) were to be extended to the right, there would be points in each subsequent wavelength (or cycle) at which $\Phi = 1.75\pi$:

in the first cycle, $\Phi = 1.75\pi$ corresponds to d
in the next cycle, $\Phi = 1.75\pi$ corresponds to $d + D$

and in subsequent wavelengths

$\Phi = 1.75\pi$ corresponds to $d + 2D, d + 3D, d + 4D \dots etc.$

Figure 17 illustrates the basic principle of E.D.M. because x (i.e. distance measured along the x-axis) can be the base-line length. If, in a base-line measurement, the reading of the phase angle (i.e. Ø) is 1.75π ($= 315°$), then the length of the base line is d or $(d + D)$ or $(d + 2D)$ or $\dots$ etc.

Since the range, D, of the instrument will be at least 1 km, the ambiguity can be eliminated by estimating the base-line length from, say, a map.

In practice, more elaborate modulation patterns may be used but the principle remains the same. The phase angle is determined at the master station after the carrier wave has passed over the base-line twice, once in each direction, and it is measured by, for example, comparing it with another signal whose phase angle can be varied in a known way. This process is called phase comparison.

SPECIMEN QUESTION 8

(a) Taking the velocity of light to be 300 000 km/s calculate the frequencies required in order that a geodimeter has three measuring units of 2.5 m, 50 m, and 1000 m. Also calculate the modulating frequencies and explain how the three scales are used to obtain an accurate value of the base-line length.

(b) Explain how the possibility of misalignment of the reflector for a geodimeter is overcome.

SOLUTION

(a) Referring to Fig. 17 the measuring unit of 2.5 m can be obtained from the basic frequency shown in (a) and it will be half the wavelength because the wave makes a double journey over the base-line.

i.e. $\qquad \lambda_1 = 5$ m and $\therefore f_1 = \dfrac{3 \times 10^8 \text{ m/s}}{5} = 60$ MHz
$\qquad\qquad\qquad\qquad\qquad\qquad$ (i.e. Megahertz)

similarly $\quad \lambda_2' = 2 \times 50$ m $\therefore f_2' = \dfrac{3 \times 10^8}{100} = 3$ MHz and $\quad \lambda'_3 = 2 \times 1000$ m $\therefore f'_3 = \dfrac{3 \times 10^8}{2000} = 150$ kHz

In the above f'_2 and f'_3 are the modulated frequencies, like Fig. 17(c). The frequencies required to produce these are f_2 and f_3 given by the following:

since $\qquad \lambda'_2 = 20\lambda_1, \qquad \dfrac{f_1}{f_2} = \dfrac{20}{21} \quad \therefore f_2 = 63$ MHz

since $\qquad \lambda'_3 = 400\lambda_1, \qquad \dfrac{f_1}{f_3} = \dfrac{400}{401} \quad \therefore f_3 = 60.15$ MHz

To illustrate how the three scales are used suppose the readings obtained are:

$\varnothing_3 = 190°$, $\varnothing_2 = 199°$ and $\varnothing_1 = 20°$.

Also, assume that from a map the base-line measures about 1.5 km then, length of base-line is

contribution from map			1000 +	metres
+ scale$_3$ = $\dfrac{190}{360} \times 1000$	= 527.7,	say,	500 +	„
+ scale$_2$ = $\dfrac{199}{360} \times$ 50	= 27.6	say,	27.5 +	„
+ scale$_1$ = $\dfrac{20}{360} \times$ 2.5 =	0.139		0.139	
Length of base-line =			1527.639 m	

Note, for example, that the reading from scale$_2$ of 27.6 is rounded down to 27.5, this being a multiple of 2.5 m, the measuring unit for the fine scale. The balance of 0.1 is ignored because the fine scale gives this more accurately as 0.139 m.

(b) A device sometimes called a "cube corner prism" is used. Its construction is indicated in Fig. 18 in which $ABCDEFG$ are the visible corners of a cube and $ABDF$ is the prism. It has the properties that:

(i) a light ray entering the face BDF, no matter what the angle of incidence, is reflected back along a parallel path; and

(ii) the internal path of any such ray in the prism always has the same length—so that corrections are easily applied for distance errors caused by the velocity of light having different values in glass and air.

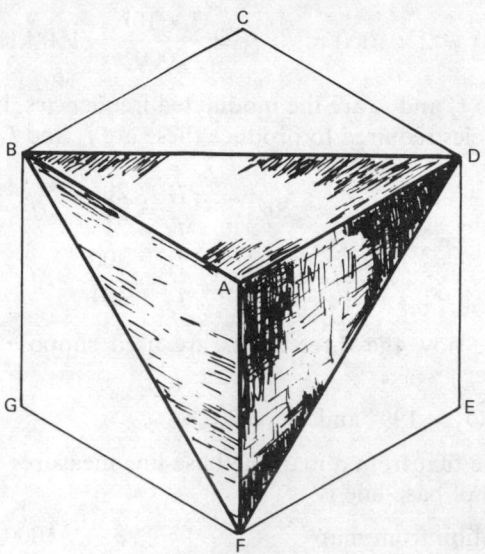

FIG. 18. *The shape of a "cube corner prism", shown shaded, and its relation to a cube.*

SPECIMEN QUESTION 9

(a) The specification of a Tellurometer reads:

$$\text{accuracy:} \quad \pm 10 \text{ mm} \pm 3 \text{ ppm}$$
$$\text{range: } 30 \text{ m to } 100 \text{ km}$$

Explain the meaning of these figures and calculate the standard deviation implied.

(b) Briefly state what corrections have to be made to a distance reading by E.D.M. in order to use it for map making.

SOLUTION

(a) Errors quoted are in the form $\pm k \pm rd$ where k represents instrument errors and rd is the operational error proportional to the measured distance, d. In this example $k = \pm 10$ mm and $r = \pm 3$ *parts per million*

$$\text{i.e. } r = \pm 3 \text{ mm/km}$$

see p. 199 for standard deviation.

when $d = 100$ km,

$$\text{s.d.} = \sqrt{10^2 + (3 \times 100)^2} = \pm 300.17 \text{ mm}$$
$$\text{say} \quad \pm 300 \text{ mm}$$

when $d = 30$ m,
$$\text{s.d.} = \sqrt{10^2 + \left(\frac{3 \times 30}{1000}\right)^2} = \pm 10 \text{ mm}$$

(b) *Corrections*

(i) A slope correction (*see* p. 20) has to be applied to compensate for the difference in level between the master and remote instruments.

(ii) The altitude correction (*see* p. 20) has to be applied to give the distance at msl.

In long sights allowance has to be made for curvature and refraction and in geodetic surveys the distances calculated have to be the values measured along great circles—i.e. the curved distances applicable in spherical triangles.

EXAMINATION QUESTIONS

1. (a) A line is measured along a long gentle slope using a 20 m chain. The gradient of the line is measured with an Abney level and found to be 5°30'. The slope distance is recorded as 150.25 m. The length of the chain was found subsequently to be only 19.85 m. Calculate the true horizontal length of the line.

(b) Assuming that a point may be plotted on a plan to ±0.2 min, calculate the maximum gradient to the nearest 0.5 degree over which no slope correction is necessary on a line the slope length of which is 150 m, when the line is to be plotted on a plan to a scale of

(i)   1:500
(ii) 1:1250

2. A line is measured using a steel band along the ground. The ground slopes evenly over the length of the line and the recorded slope distance is 287.346 m. The band is standardised at 20° C and after measuring the line, the length of the band was checked at this temperature and found to measure 30.021 m, though it had been thought to measure 30 m exactly. The reduced levels of the two ends of the line are 386.34 m and 370.41 m above datum and the temperature at the time the line was measured was 15° C. If the coefficient of linear expansion of steel is 0.00001125 per degree C, calculate the true horizontal length of the line reduced to datum level and corrected for temperature and standardisation. The mean radius of the Earth may be taken as 6370 km.

3. Two gauge marks are inscribed on one of the flat horizontal girders of a girder bridge for the purpose of checking the amount of expansion or contraction of the bridge. During one such measurement a steel tape was laid flat along the girder under a pull of 45 N at a temperature of 30° C and the observed length was 29.980 m.

Calculate the actual distance between the points (to the nearest millimetre) if the 30-m tape was standardised at 15° C under a pull of 145 N and hanging in catenary. Weight of tape was 7.6 N, coefficient of expansion 0.000 012 per °C, density of steel 7860 kg/m^3 and Young's modulus for steel 2×10^{11} N/m^2.

4. The following data relate to an etched steel tape of 30-m nominal length.

Section: 12.5 mm × 0.25 mm
Mass: 0.74 kg
Coefficient of linear expansion: 0.000 011 per °C
Young's modulus of elasticity: 200×10^9 N/m^2

If the tape was standardised "flat" under a pull of 45 N at a temperature of 20° C show that, when used in catenary at 20° C, a pull of 128 N is required to bring the tape to standard length (measured to the nearest millimetre).

5. A works base-line measured 29.860 m on a steel tape supported on three equidistant supports, but hanging freely between them. The temperature was 22° C, the pull was 90 N and the reduced levels of the supports were, from one end to the other, 1602.5, 1603.5 and 1602.0 m above Ordnance Datum.

If the 30-m tape used was standardised in catenary under 90 N pull at 7° C calculate to the nearest millimetre the true length of the base-line reduced to mean sea level. The weight of the tape was 12 N, the coefficient of linear expansion of steel = 0.000 011 per °C and the radius of the Earth is 6370 km.

6. The following readings were taken in measuring a base-line with a steel tape suspended in catenary in five spans.

Span	Mean reading of tape (m)	Difference in level between index marks (m)	Tension (N)	Mean temperature (°C)
1	100.155	3.1	110	20
2	100.140	0.9	220	23
3	100.060	1.2	110	25
4	100.108	3.1	110	27
5	100.182	2.0	110	27

The tape reading was 100.005 m when calibrated in catenary under a tension of 110 N and at a temperature of 18° C between two points at the same level precisely 100 m apart.

Other tape constants are: width of tape = 3 mm; thickness of tape = 0.625 mm; density of steel = 7860 kg/m^3; E for steel = 200×10^9

N/m^2; coefficient of expansion of steel = 0.000 011 per °C. Compute the length of the base-line.

7. Recent developments in optical and electromagnetic distance measuring instruments have produced new possibilities in surveying techniques. Discuss the impact which these developments have made, referring to the major changes which have taken place in surveying methodology.

8. (a) A steel tape weighing 0.2 N per metre is suspended between two pulleys at the same level. It carries two marks which are 29.990 m apart horizontally when the tensioning masses are 7 kg. How far apart would the two marks be if the masses were increased to 14 kg?

$$\text{Density of tape material} = 7500 \text{ kg/m}^3$$
$$\text{Modulus of elasticity of tape material} = 205 \text{ GN/m}^2$$

(b) Which methods, other than catenary measurement, would you suggest for measuring, with very high accuracy, the lengths of three lines whose approximate lengths were 1 km, 50 km and 500 km? State the principles on which the methods are based.

Chapter 3

USE OF THE LEVEL

FIELDWORK

In surveying the term levelling is used to describe the process by which the relative heights of a number of points are determined. In Britain the reduced levels of the points are usually referred to Ordnance Datum or, occasionally, to an arbitrary datum. Thus, we say point X is 106.00 m A.O.D. (i.e. above Ordnance Datum) and point Y is 112.00 m A.O.D. in preference to saying, merely, that Y is 6 m above X.

The equipment needed for levelling is a dumpy level, or a tilting level, and a levelling staff, which, usually, is a wooden staff calibrated in metres and decimals of a metre. The erstwhile common Sopwith staff, which was calibrated in Imperial measure, is being replaced by staffs up to 4 m or 5 m in length which can be read direct to either the nearest 5 mm or 10 mm—and the student should familiarise himself with the pattern available at his school or office.

The following notes refer to Fig. 19, which is a sketch of a simple level run.

The level is set up at position 1, as described on p. 4 (and in Fig. 3), so that the line of sight through the telescope always lies on the same horizontal plane.

The relative levels of A, B, C and D can be determined by holding the staff at each in turn and reading the height at which the line of sight intercepts it. For example, if the staff reading at A is 3 m and at B is 1.8 m, then B is 1.2 m higher than A.

The point A is a bench mark (B.M.). Ordnance surveyors have established bench marks at frequent intervals over the whole country; the reduced levels of these marks, reckoned from Ordnance Datum, are known and can be found on Ordnance Survey maps. By starting the level run at a bench mark, the reduced levels of all subsequent points can be determined. For example, if the staff readings taken from position 1, at A, B, C and D, are 3 m, 1.8 m, 3.5 m and 1 m respectively, and if the reduced level (R.L.) of A is 32 m A.O.D., then the reduced levels of B, C and D are respectively 33.2 m, 31.5 m and 34 m A.O.D.

When an instrument is set up in a new position for levelling the first staff reading is called a backsight. The last staff reading before moving

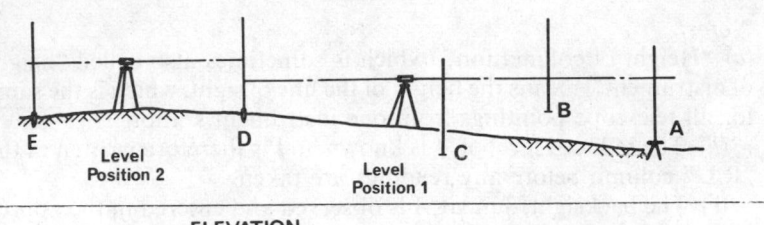

ELEVATION

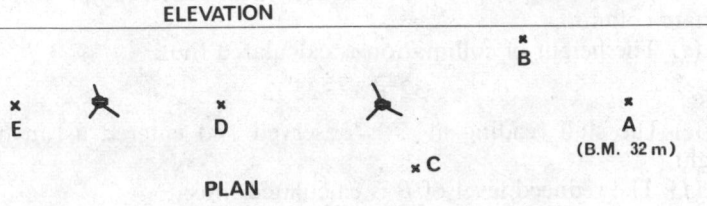

PLAN

FIG. 19. *Use of levelling staffs.*

the instrument is a foresight, and all other staff readings are called intermediate sights. Thus, from position 1 the reading at A is a backsight, the reading at D is a foresight and those at B and C are "inter sights" (a common abbreviation for "intermediate sights"). When the instrument is moved to position 2 the new reading, 2 m say, at D is a backsight and, assuming the level run ends at E, the reading at E, 1.8 m say, is a foresight. The point D is called a change point (C.P.), the staff being kept there while the position of the instrument is changed.

BOOKING

The terms defined above are used when booking the field observations by either of the two available methods. The layout for each method must be learned, and so must the arithmetic checks.

The staff readings quoted in the previous section are booked in the same way for both methods, but the reduced levels are calculated differently in each method.

1. Height of collimation method

Back-sight	Inter sight	Fore-sight	Height of collimation	R.L.	Distance	Remarks
3.0			35.0	32.0		A, B.M. 32.0 m A.O.D.
	1.8			33.2		B
	3.5			31.5		C
2.0		1.0	36.0	34.0		D, C.P.
		1.8		34.2		E

Notes

(*a*) "Height of collimation" (which is sometimes also called "height of instrument") means the height of the line of sight, which is the same for all telescope pointings from one instrument station.

(*b*) The reduced level of *A* is known and is therefore written in the "R.L." column before any readings are taken.

(*c*) The backsight, 3 m, at *A* is observed and entered in the appropriate column.

(*d*) The height of collimation is calculated thus:

$$32 + 3 = 35 \text{ m}$$

(*e*) The staff reading at *B* is observed and entered as an inter sight.

(*f*) The reduced level of *B* is calculated thus:

$$35.0 - 1.8 = 33.2 \text{ m}$$

(*g*) The reduced levels of *C* and *D* are similarly obtained.

(*h*) *D* is a change point. While the staff is held at *D* the instrument is moved to position 2, where the new line of sight is at a different level from that of position 1. It is calculated as before, namely:

$$34 + 2 = 36 \text{ m, the new height of collimation}$$

(*i*) The reduced level of *E* is calculated thus:

$$36.0 - 1.8 = 34.2 \text{ m}$$

(*j*) The arithmetic is checked by applying the following formula:

$$\frac{\text{Sum of}}{\text{backsights}} - \frac{\text{Sum of}}{\text{foresights}} = \text{Last R.L.} - \text{First R.L.}$$

i.e., with the figures tabulated above,

$$5.0 - 2.8 = 34.2 - 32.0$$

The "Distance" column is sometimes used to describe the staff positions (*see* Specimen Question 10).

2. Rise and fall method

Back-sight	Inter sight	Fore-sight	Rise	Fall	R.L.	Distance	Remarks
3.0					32.0		*A*, B.M. 32.0 m A.O.D.
	1.8		1.2		33.2		*B*
	3.5			1.7	31.5		*C*
2.0		1.0	2.5		34.0		*D*, C.P.
		1.8	0.2		34.2		*E*

Notes

(*a*) The staff readings and R.L.s are booked in the same order as in the previous method.

(*b*) The staff reading at *A* is 1.2 m greater than at *B*; evidently the foot of the staff has risen 1.2 m in the move from *A* to *B*. There is thus a rise of 1.2 m and, consequently,

$$\text{R.L. of } B = 32.0 + 1.2 = 33.2 \text{ m}$$

(*c*) Similarly, there is a fall from *B* to *C* of $3.5 - 1.8 = 1.7$ m.

hence $\qquad$ R.L. of $C = 33.2 - 1.7 = 31.5$ m

(*d*) The reduced level of any staff station is obtained by adding the "Rise" to, or subtracting the "Fall" from, the reduced level of the previous staff station.

(*e*) By this method there is an additional arithmetic check thus:

$$\frac{\text{Sum of}}{\text{backsights}} - \frac{\text{Sum of}}{\text{foresights}} = \frac{\text{Sum of}}{\text{rises}} - \frac{\text{Sum of}}{\text{falls}} = \text{Last R.L.} - \text{First R.L.}$$

i.e., with the figures given above,

$$5.0 - 2.8 = 3.9 - 1.7 = 34.2 - 32.0$$

Rule

In both methods the check on arithmetic works only if the first staff reading booked is a backsight and the last staff reading booked is a foresight. This check should be carried out at the bottom of every page of levels (*see* Specimen Question 11).

Comparison of methods

In the rise and fall method each reduced level is calculated from the previous reduced level, and therefore a satisfactory arithmetic check ensures that all the reduced levels are correctly calculated. In the height of collimation method mistakes in isolated reduced levels can occur without upsetting the check. On the other hand, this method involves less arithmetic than the rise and fall method, and is therefore quicker.

SPECIMEN QUESTION 10

A site for a new road across a filled-in valley has been prepared, and centre line pegs at 50-m intervals have been levelled to indicate a slope of one in a hundred. The first and last stations, chainages 0 and 500 m, are on bed rock, but it is feared that there has been subsidence in between and the pegs were re-levelled. From the following data determine the reduced levels of the pegs and indicate whether there has been any subsidence and if so, how much.

Instrument station	Staff readings (Nos. in brackets indicate chainage of pegs)		
a	0.620 (B.M)	2.120 (0)	1.625 (50 m)
b	1.305 (50 m)	0.810 (100 m)	0.320 (150 m)
c	2.075 (150 m)	1.585 (200 m)	1.095 (250 m)
d	2.710 (250 m)	2.205 (300 m)	1.695 (350 m)
e	2.910 (350 m)	2.400 (400 m)	1.890 (450 m)
f	1.470 (450 m)	0.965 (500 m)	

SOLUTION

Back-sight	Inter sight	Fore-sight	Ht. of coll.	R.L.	Dist. (m)	Re-marks	Design R.L.	Sub-sidence
						B.M.		
0.620			10.620	10.000		(10.000)		
	2.120			8.500	0		8.500	
1.305		1.625	10.300	8.995	50	C.P.	9.000	0.005
	0.810			9.490	100		9.500	0.010
2.075		0.320	12.055	9.980	150	C.P.	10.000	0.020
	1.585			10.470	200		10.500	0.030
2.710		1.095	13.670	10.960	250	C.P.	11.000	0.040
	2.205			11.465	300		11.500	0.035
2.910		1.695	14.885	11.975	350	C.P.	12.000	0.025
	2.400			12.485	400		12.500	0.015
1.470		1.890	14.465	12.995	450	C.P.	13.000	0.005
		0.965		13.500	500		13.500	
11.090		7.590						

CHECK
Sum of backsights − Sum of foresights = 11.090 − 7.590 = 3.500 m
Last R.L.   − First R.L.   = 13.500 − 10.000 = 3.500 m

NOTES
(i) The reader should, as an exercise, recalculate the reduced levels by the rise and fall method.

(ii) The "Design R.L." figures are calculated from the given gradient of 1/100. Thus at chainage (50 m) the embankment should be 0.50 m higher than at chainage (0).

(iii) The subsidence is calculated by subtracting the figures in the column "R.L." from the corresponding figures in the column "Design R.L."

(*iv*) The change points in the question are those chainages at which two staff readings are given.

(*v*) It is an instructive exercise to draw to scale a section through the embankment showing staff positions, lines of sight and possible instrument stations. Suitable scales are 5000 to 1 horizontally and 100 to 1 vertically. These scales will not show the subsidence, but the exercise will test the reader's understanding of the levelling procedure.

LONG SIGHTS

The earth is approximately spherical in shape, and lines of constant altitude are therefore circles with their centres at the centre of the Earth. If the lines of sight were straight it is easily shown (*see* p. 96) that the staff reading would be too great by 78 D^2 mm, D being the distance in km between the instrument and the staff. 78 D^2 mm is the error due to curvature.

It is well known that the density of the Earth's atmosphere varies, the air being rarer at greater altitudes. Owing to this phenomenon, light rays through the atmosphere are refracted and, as they are curved towards the Earth's surface, it partly offsets the error due to curvature. On average, this reduction is about 14%, so that the error due to curvature and refraction is usually taken as 67 D^2 mm. The corrected staff reading is obtained by deducting this amount from the observed staff reading.

LEVELLING PROCEDURE

Errors in reduced levels may be caused by (*i*) curvature and refraction, (*ii*) collimation error, (*iii*) carelessness in holding the staff, reading the staff, setting up the instrument or booking, and can be eliminated—or, at worst, detected—by good levelling practice:

(*a*) by keeping backsights and foresights of equal length, which eliminates errors (*i*) and (*ii*);

(*b*) by using a staff level, a device which when held against the staff indicates whether or not it is vertical;

(*c*) by using one of the recognised methods of booking, with checks to detect arithmetical mistakes;

(*d*) by always commencing the level run at a bench mark and ending it at a bench mark, in order to discover any closing error.

SPECIMEN QUESTION 11

(*a*) The groups of figures below refer to staff readings taken with a level from instrument stations, *A*, *B*, *C*, *D* and *E*. The first and last readings in each group are the backsight and foresight respectively.

The backsight from station *A* was taken with the staff held on a bench mark at 204.110 m above Ordnance Datum.

A: 2.680, 0.875, 0.980, 0.430;　　*B:* 1.665, 1.440, 0.625
C: 1.010, 1.690, 1.225　　　　　　*D:* 2.445, 3.575, 3.880, 2.880
　　　　　　　E: 2.735, 2.005, 2.390

Book the readings by the height of instrument method and determine the reduced level of each staff station.

Assume that the reading 3.880 from station *D* comes at the bottom of the page in the levelling book and leave a space in your booking to indicate the turn over the page. Make all the checks you think necessary.

(*b*) Indicate the main differences in instruments used, and in field procedure, between ordinary levelling and precise levelling. Assume in the former case an allowable discrepancy in mm of $13\sqrt{D}$ and in the latter case $2.4\sqrt{D}$, where *D* is the length of the circuit in km.

SOLUTION

(*a*) Although the question asks for the height of collimation method of booking, the solution given below is by the rise and fall method. An examination candidate must not ignore instructions like this, of course, for all his work by the wrong method would presumably yield no marks. The levels are here reduced by the rise and fall method however to demonstrate the method, and the reader should

	Back-sight	Inter-sight	Fore-sight	Rise	Fall	Reduced level	Remarks
	2.680					204.110	B.M. 204.110 m A.O.D.
		0.875		1.805		205.915	
		0.980			0.105	205.810	
	1.665		0.430	0.550		206.360	C.P.
		1.440		0.225		206.585	
	1.010		0.625	0.815		207.400	C.P.
		1.690			0.680	206.720	
	2.455		1.225	0.465		207.185	C.P.
		3.575			1.120	206.065	
		3.880			0.305	205.760	
Totals	7.810		2.280	3.860	2.210		
Add last inter sight		3.880					
		6.160					

work through the solution again using the required height of instrument method.

CHECK

CHECK
   Sum of backsights − Sum of foresights =   7.810 −   6.160 = 1.65 m
   Sum of rises − Sum of falls   =   3.860 −   2.210 = 1.65 m
   Last R.L. − First R.L.   = 205.760 − 204.110 = 1.65 m

	Back-sight	Inter sight	Fore-sight	Rise	Fall	Reduced level	Remarks
		3.880				205.760	
	2.735		2.880	1.000		206.760	C.P.
		2.005		0.730		207.490	
			2.390		0.385	207.105	
Totals	2.735		5.270	1.730	0.385		
Add	3.880						
	6.615						

CHECK
   Sum of backsights − Sum of foresights =   6.615 −   5.270 = 1.345 m
   Sum of rises − Sum of falls   =   1.730 −   0.385 = 1.345 m
   Last R.L. − First R.L.   = 207.105 − 205.760 = 1.345 m

It should be noticed how the intermediate sight, 3.880, is included once with the backsights and once with the foresights for the purpose of applying the check, in accordance with the rule on p. 39.

(b) The main differences between instruments used are given on pp. 3–6. The main points to be observed in ordinary levelling procedure are listed on p. 41. In precise levelling the procedure is refined by:

   (i) using a superior staff;
   (ii) using a superior instrument with a parallel plate micrometer, reversible telescope and taking each staff reading twice, once with the telescope in its normal position and once inverted;
   (iii) correcting for curvature and refraction whenever backsight and foresight are not of equal length;
   (iv) protecting the instrument from the sun with an umbrella;
   (v) correcting the staff readings for temperature; and
   (vi) frequent tests of the equipment.

RECIPROCAL LEVELLING

Where a line of levels is to cross a wide river, for example, it may not be possible to make backsights and foresights equal. In this case instrument errors and the effects of curvature and refraction can be eliminated by reciprocal levelling (*see* Fig. 20). The procedure is as follows:

(*a*) Staff stations are established at A and B on either side of the river.

(*b*) The level is set up as near A as possible (but far enough away

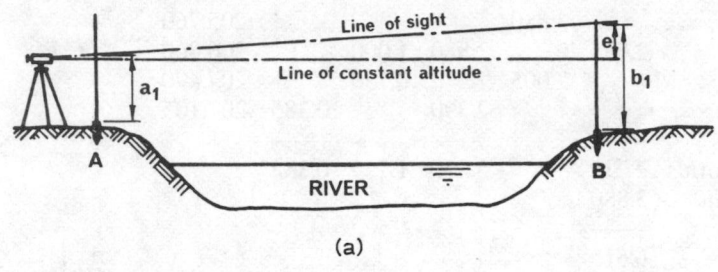

(a)

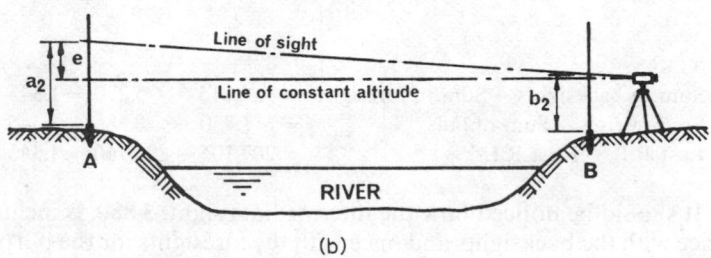

(b)

FIG. 20. *Reciprocal levelling.*

to focus it clearly), Fig. 20 (*a*). Staff readings a_1 and b_1 are taken on a staff held at A and B in turn. Then

$$\text{Apparent difference in level} = b_1 - a_1, \; A \text{ above } B$$

If $e =$ (unknown) error in length AB, then

$$\text{True difference in level} = b_1 - e - a_1, \; A \text{ above } B$$

(*c*) *The instrument* is then set up as near as possible to B (*see* Fig. 20 (*b*)). Then since e is the same as in (*b*),

$$\text{True difference in level} = b_2 - (a_2 - e)$$
$$= b_2 - a_2 + e, \; A \text{ above } B$$

As before,

Apparent difference in level $= b_2 - a_2$, A above B

Combining all these results,

True difference in level $= \frac{1}{2}[(b_1 - e - a_1) + (b_2 - a_2 + e)]$

A above B

$\qquad = \frac{1}{2}[(b_1 - a_1) + (b_2 - a_2)]$, A above B

Hence the rule for reciprocal levelling is: true difference in level is equal to half the sum of the apparent differences in level.

It should be noticed that, in the above proof, e may be due to a collimation error, or it may be the error due to curvature and refraction. It follows that the result is true in all circumstances.

SPECIMEN QUESTION 12

In what circumstances is reciprocal levelling carried out and what are the advantages of the method?

Two points X and Y are 700 m apart. A level is set up at X and Y in turn and the following readings taken:

Level at X. Height of collimation 1.650 m. Reading of staff on Y 1.635 m.

Level at Y. Height of collimation 1.632 m. Reading of staff on X 1.625 m.

Calculate: (*a*) the reduced level of Y if the reduced level of X is 5.040 m A.O.D., and (*b*) the collimation error of the instrument.

SOLUTION

The first part is answered above.

In the second part the procedure differs from that described above in so far as the level is set up exactly over the peg in each case, so that the height of collimation in the question replaces the near staff reading in the text.

(*a*) From X, apparent difference of level $= 1.650 - 1.635 = 0.015$ m, Y above X. From Y, apparent difference of level $= 1.625 - 1.632 = -0.007$ m, Y above X.

$\therefore$ half the sum of the apparent differences of level is:

$$\tfrac{1}{2}(0.015 - 0.007) = 0.004 \text{ m}$$

hence $\qquad$ R.L. of $Y = 5.040 + 0.004 = \underline{\underline{5.044 \text{ m, A.O.D.}}}$

(*b*) Error due to curvature and refraction is

$$67\ D^2 \text{ mm} = 67 \times (0.7)^2 = 33 \text{ mm}$$

i.e. for each instrument position the distant staff reading is 33 mm too high.

Let   collimation error = e m in 700 m, up

Then, with the instrument at X, the distant staff reading, corrected for curvature, refraction and collimation error, would be $1.635 - 0.033 - e$.

but   height of collimation = $5.040 + 1.650 = 6.690$ m
∴   true level of Y = $6.690 - (1.635 - 0.033 - e)$
                                = 5.044, from part (a) above.
hence $6.690 - 1.635 + 0.033 + e = 5.044$
                         ∴ $e = -0.044$ m

i.e.   Collimation error = 0.044 m down in 700 m

NOTES

(*i*) In questions such as this the collimation error must be specified completely. This entails stating the error in the staff reading (0.044 m), the distance over which the error occurs (700 m) and the "direction" of the error, up or down. Alternatively, it can be described as an angle.

(*ii*) A collimation error is said to be up if it causes an increase in the staff reading.

TACHEOMETRY PRINCIPLES USED IN SURVEYING

In many levels the telescope is fitted with stadia lines. Then, the distance to the staff from the instrument is given by the product, staff intercept × the multiplying constant. The staff intercept is the length of staff appearing between the stadia lines, and the multiplying constant is usually 100.

The tacheometry principle is explained at greater length in Chapter 5, commencing on p. 69.

SPECIMEN QUESTION 13

P and Q are two points on opposite banks of a river about 100 m wide. A level with an anallatic (*see* p. 72) telescope and a constant of 100 is set up at A on the line QP produced, then at B on the line PQ produced and the following readings taken on to a graduated staff held vertically at P and Q:

| From | To | Staff readings in metres | | |
		Upper stadia	Collimation	Lower stadia
A	P	1.565	1.425	1.285
	Q	1.000	0.370	Below ground
B	P	3.240	2.595	1.950
	Q	1.605	1.450	1.295

What is the true difference in level between P and Q and what is the collimation error of the level expressed in seconds of arc, there being 206 265 seconds in a radian?

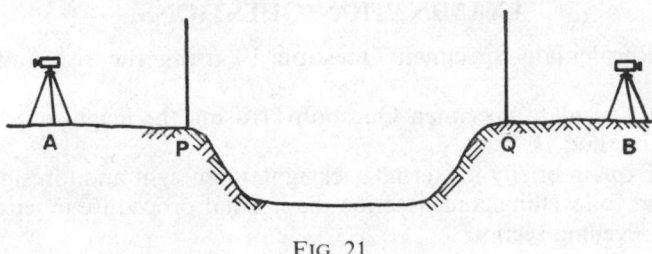

FIG. 21.

SOLUTION

Referring to Fig. 21,

$$AP = (1.565 - 1.285) \times 100 = 28 \text{ m}$$
$$AQ = 2(1.000 - 0.370) \times 100 = 126 \text{ m}$$
$$BP = (3.240 - 1.950) \times 100 = 129 \text{ m}$$
$$BQ = (1.605 - 1.295) \times 100 = 31 \text{ m}$$

With instrument at A, apparent difference of level is

$$1.425 - 0.370 = 1.055 \text{ m, } Q \text{ above } P$$

With instrument at B, apparent difference of level is

$$2.595 - 1.450 = 1.145 \text{ m, } Q \text{ above } P$$

True difference of level $= \frac{1}{2}(1.055 + 1.145)$, Q above P
$$= 1.100 \text{ m } Q \text{ above } P$$

Let collimation error be e metres up in 100 m, then
error in $AP = 0.28e$, and error in $AQ = 1.26e$
True difference of level $= 1.100 = (1.425 - 0.28e) - (0.370 - 1.26e)$
$$\therefore 1.100 = 1.055 + 0.98e$$

hence
$$e = 0.045/0.98 \text{ in 100 m, up}$$
$$= 0.045/98$$
$$= 0.045/98 \times 206\,265 = \underline{\underline{95 \text{ seconds up}}}$$

NOTE

Problems of this nature can be solved by the above method only if the error due to curvature and refraction is negligible on all sights, for this error depends on the square of the distance. In this question the error due to curvature and refraction in the longest sight, BP, is

$$67 \times 0.129^2 \simeq 1 \text{ mm}$$

Hence the error due to curvature and refraction for all sights can be ignored, since staff readings here are taken only to the nearest 5 mm.

EXAMINATION QUESTIONS

1. Recalculate Specimen Question 10 using the rise and fall method.

2. Recalculate Specimen Question 11 using the height of instrument method.

3. Explain briefly the terms backsight, inter sight and foresight as applied to levelling, and describe the normal procedure in carrying out a levelling section.

A levelling section is run downhill from A to B and is linked to Ordnance B.M.s at each end. The staff readings taken in order are as follows, the change points being all on the section: 2.67 (O.B.M. 78.800), 0.53 (station A), 1.11, 1.87, 2.57, 3.92, 1.02, 2.15, 3.36, 3.80, 1.09, 2.70, 3.61, 3.94 (station B), 1.98, 2.08, 1.73, 1.30, 2.66, 1.61 (O.B.M. 73.295).

If the staff points on the section are 15 m apart, draw up a level book page showing these readings, obtain the R.L.s and apply the usual checks.

HINT:
Since the section is downhill a sudden decrease in staff reading (e.g. 3.92, 1.02) signifies a change point; after station B all the readings are foresights and backsights taken for the purpose of checking on to the O.B.M.

4. A dumpy level is used to level up a slope of 1 in 6 from A to B, the instrument axis being approx. 1.50 m above ground at each set up. The following readings are obtained:

Backsight	Foresight	Remarks
3.975		Pt. A O.B.M. 135.605
3.660	0.310	C.P. 1
3.995	0.365	C.P. 2
3.105	0.455	C.P. 3
3.910	0.670	C.P. 4
	0.505	Pt. B

Draw up a suitable form of booking and calculate the reduced level of B.

In order to check the line of collimation, the level is then set up midway between two pegs X and Y, 70 m apart on fairly level ground, and, with the bubble central, sights are taken on a levelling staff held on X and Y in turn; these readings being 2.000 and 1.355 respectively. The level is then set up near X and, with the bubble central, staff readings are taken on X and Y; these being 1.500 and 0.875. Find the collimation error, estimate the error in the reduced level of B and hence find its true level.

5. A level was set up midway between two staff stations X and Y. The bubble was brought to the centre of its run by means of the foot-screws when sighting on X and sighting on Y. The readings on X and Y were X_1 and Y_1. The instrument was then moved to Z, which was in line with X and Y, but not between them, so that the distance ZY was 11 times the distance ZX, and the readings with the bubble central were X_2 and Y_2.

Deduce an expression for the true staff reading on Y when the instrument is adjusted at Z. How would you carry out the necessary adjustment?

6. Levelling checks carried out on the invert of a large brick sewer laid to a uniform gradient between two manholes A and B gave the following staff readings:

Backsight on T.B.M. 153.08	0.63
Intermediate sight at A	1.41
Intermediate sight at $A + 10$ m	1.47
Foresight at $A + 20$ m	1.55
Backsight at $A + 20$ m	0.83
Intermediate sight at $A + 30$ m	0.89
Intermediate sight at $A + 40$ m	0.93
Foresight at B	0.99

Reduce the levels, applying the normal checks, and determine any discrepancy in the gradient between A and B which are 50 m apart.

7. (a) The following readings were taken, in the order given, with a level and vertical staff. Draw up a level book page for the readings and reduce the levels, applying all the usual arithmetical checks.

0.716 O.B.M. (38.330 m A.O.D.); 1.119, 1.925, 2.400, (2.150 and 1.290) C.P., 1.662, 2.547, (2.821 and 3.263) C.P., 2.477, 2.074, 1.842 T.B.M.

(b) What error would occur in the level of the T.B.M. if the staff had been tilted back at an angle of $4°$ to the vertical for all readings?

(c) Comment upon the two possible methods of reducing the levels, indicating the advantages of each.

8. Given below are the field notes made during a levelling for the construction of a road. Certain figures have become obliterated from the booking. However, sufficient data remains to make it possible to complete the booking. Copy the figures to your answer and insert the

missing values. Then calculate the gradient from A to C which is a straight line.

Back-sight	Inter-mediate sight	Fore-sight	Rise	Fall	Reduced Level (m)	Dist. (m)	Remarks
2.841					60.000	0	At A
			0.236		60.236	15.00	
	2.872					30.00	
			0.553			45.00	
		2.810		0.491		60.00	
	1.021			0.115		75.00	At B
	0.378					90.00	
				1.599		105.00	
1.426		2.773		0.796	58.164	120.00	
			0.689		58.853	135.00	
	2.103					150.00	
			0.866			165.00	At C
5.173		6.820					

CHAPTER 4

THE USE OF THE THEODOLITE

I. IN TRAVERSING

In all construction projects, before engineering work is commenced, pegs are placed on the site to mark important centre-lines and other key points such as the corners of buildings. Such work is named setting out, and is often performed by tapes and chains. When greater accuracy is required a theodolite may be used. Sometimes, for example when setting out curves for a road or railway, special techniques are employed which are described elsewhere in this book; otherwise, setting out by theodolite usually entails aligning a number of pegs or turning off right-angles. Apart from setting out, the theodolite finds most of its work in traversing and tacheometry, and most of Chapters 4 and 5 are devoted to these branches of surveying.

TRAVERSING

Traversing is the name given to the type of surveying operation in which a horizontal angle at each instrument station is measured by theodolite (or compass, etc.) and the horizontal distances between stations are measured independently, usually by chain. Traverses may be classed as closed or open (*see* Fig. 22). Closed traverses are obviously self-checking, for when plotted the first and last points must coincide. Open traverses can be made self-checking by commencing them at a known point and terminating them at another.

Three methods of traversing are given below, with notes on their advantages and disadvantages.

1. The method of included angles

This method can be used only if the traverse forms a closed polygon, as in Fig. 22 (*b*). Usually a bearing of one line is taken so that the traverse may be orientated.

Usually the theodolite is moved from station to station in a counter-clockwise direction round the traverse. Then, when the theodolite is at *C*, say, *B* is the backward station (i.e. the previous instrument station) and *D* is the forward station. Because the angle on the horizontal protractor increases in the clockwise direction, the natural procedure is to sight *B* first, *D* second and subtract the first reading from the second, thus calculating the internal angle *BCD*. Moving the theodolite round the traverse in a clockwise direction reverses the procedure and yields the external angles.

By sketching the traverse, joining the station points to any internal point by straight lines and summing the angles of the triangles so formed, it is easily proved that the sum of the included angles of a traverse is $(2N - 4)$ right-angles, N being the number of sides of the traverse.

In practice, the sum of the measured angles usually differs from $(2N - 4) \times 90°$ by a small amount called the closing error. This should be distributed as equally as possible between all the measured angles, but only to the accuracy of the instrument. For example, if a theodolite which is accurate to the nearest 20 seconds is set up at each of seven stations and the closing error is 2 minutes, then the values of six of the angles would each be adjusted by 20 seconds to make the traverse close.

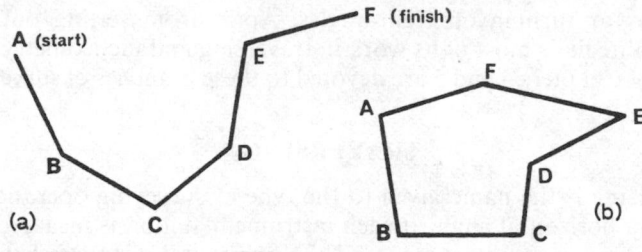

FIG. 22. (a) *Open traverse* (b) *Closed traverse.*

2. The loose (or free) needle method

Referring to Fig. 22 (*a*), the instrument would be set up at *A* and the magnetic bearing of the line *AB* recorded. Thereafter it would be set up at *B*, *C*, *D* and *E* in turn and the magnetic bearings of the two lines which meet at each point recorded. From *F* the magnetic bearing of *EF* would be observed. This procedure can be shortened by setting up only at every other instrument station, but the advantage of the method described is that it gives a necessary check on the magnetic bearing of each line (*see* Specimen Question 14).

It should be realised that on a theodolite capable of reading angles direct to 20 seconds the compass can seldom be read to a better accuracy than the nearest 5 minutes, if that. The accuracy of the loose needle method is therefore compass accuracy and not theodolite accuracy. In fact, prismatic compasses or dials are instruments more suited to the method.

3. The fast needle method

Referring again to Fig. 22 (*a*), the theodolite is set up at *A*. With the horizontal vernier reading zero, the telescope is pointed to magnetic north. The lower clamp is fixed, the upper clamp is freed and the telescope is pointed to *B* so that the horizontal scale now reads α, say, the magnetic bearing of *AB*.

With the horizontal scale still reading α the theodolite is moved to *B* and set up there without freeing the upper clamp. The telescope is sighted back to *A*, the lower clamp is tightened and the telescope is transitted. Only then is the upper clamp freed. Consequently, when the telescope is pointed to *C* the horizontal scale reading is then the magnetic bearing of *BC*.

The procedure of setting up, pointing back to the previous station, transitting and only then freeing the upper clamp is the same at each subsequent station and gives the magnetic bearing of each line in turn. Although the magnetic bearing of any particular line is only as accurate as the compass will allow, the difference between bearings of adjacent lines is obtained with theodolite accuracy.

SPECIMEN QUESTION 14

A straight tunnel is to run between two points *A* and *B* whose co-ordinates, together with those of a third point *C*, are given below.

	Northings	Eastings
A	0 m	0 m
B	3014	256
C	1764	1398

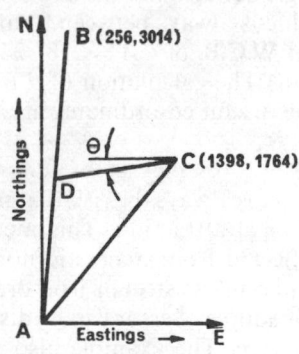

FIG. 23.

A shaft is to be sunk at *D*, the mid-point of *AB*, but it is impossible to measure along *AB* directly, so *D* is to be fixed from *C*.

Calculate (*a*) the co-ordinates of *D*, (*b*) the length and bearing of *CD* and (*c*) the angle *ACD*, given that the bearing of *AC* is 38° 24′ east of north.

SOLUTION

(a) Since D is the mid-point of AB, at D:

$$\text{Northing} = \tfrac{1}{2}(0 + 3014) = 1507 \text{ m}$$
and $\qquad\qquad$ $\text{Easting} = \tfrac{1}{2}(0 + 256) = 128 \text{ m}$

$\therefore$ co-ordinates of D are: 1507 m N, 128 m E.

(b) From the co-ordinates of C and D,

$$\tan \theta = (1764 - 1507)/(1398 - 128) = 257/1270$$

$\therefore$ by calculator, $\theta = 11° \ 26'$
$\therefore$ W.C.B. of D from C is $270° - \theta°$

$$= 270° - 11° \ 26' = 258° \ 34'$$

Length $CD = \sqrt{(1398 - 128)^2 + (1764 - 1507)^2} = 1295.7$ m

(c) Since W.C.B. of AC = angle $NAC = 38° \ 24'$,
$\qquad$ angle $ACD = (90° - 38° \ 24') - \theta$
$\qquad\qquad\qquad\quad = 90° - 38° \ 24' - 11° \ 26' = 40° \ 10'$

NOTE

(i) The whole circle bearing (abbreviated to W.C.B.) of a line is the angle between the line and some fixed direction, usually true north. Its value is always between $0°$ and $360°$. Thus, W.C.B. of $AC = 38° \ 24'$, but W.C.B. of $CA = 218° \ 24'$.

(ii) The calculation of θ from the expression for $\tan \theta$ illustrates how useful co-ordinates are in traverse computations.

Local attraction

There is always the risk that magnetic bearings may be in error owing to local attraction. This means that the compass needle has been deflected from magnetic north by some nearby mass of magnetic material, e.g. steel or iron ore. The following example indicates how this source of error can be discovered and allowed for in a free needle traverse. The example also shows one method of distributing the closing error.

SPECIMEN QUESTION 15

The following data refer to a small reconnaissance compass traverse made through the rocky country. Plot the traverse to a scale of 4 paces to the millimetre on squared paper, having first meaned the magnetic bearings and adjusted them, where necessary, for any discrepancy in bearing due to local attraction.

Adjust the traverse graphically and state by how much and in what direction the corrected point F differs from its plotted position.

Side	Distance	Magnetic bearings			
AB	300 paces	AB	69°	BA	248°
BC	200 „	BC	56°	CB	232°
CD	400 „	CD	121°	DC	305°
DE	150 „	DE	180°	ED	360°
EF	350 „	EF	265°	FE	89°
FG	200 „	FG	313°	GF	129°
GH	150 „	GH	246°	HG	68°
HA	200 „	HA	303°	AH	121°

SOLUTION

Side	Distance	Mean magnetic bearing
AB	300 paces	$68\frac{1}{2}°$
BC	200 „	56°
CD	400 „	125°
DE	150 „	180°
EF	350 „	265°
FG	200 „	309°
GH	150 „	246°
HA	200 „	301°
Total:	1950 „	

NOTES

(i) The magnetic bearings are examined to determine for which lines the forward and back bearings differ by exactly 180°. In this example the bearings of DE and ED are 180° and 360° respectively and it is therefore deduced that there is no local attraction at stations D or E.

(ii) Since there is no local attraction at E, for line EF, bearing $EF = 265°$ must be correct and therefore bearing $FE = 265° - 180° = 85°$. Local attraction at F is, therefore, $85° - 89° = -4°$.

(iii) The local attraction at F affects bearing FG which should be $313° - 4° = 309°$, and this figure is consistent with the back bearing of 129° for GF, there being 180° difference between these figures.

(iv) Consequently there can be no local attraction at G, so bearing $GH = 246°$ is correct.

(v) By the same kind of argument as used above it can be deduced that there is local attraction of $-2°$ at H but none at A.

(vi) Similarly, correct bearings are determined for CD and BC and, as no local attraction is evident at A or B, the true bearing here is taken as the mean of 69° and (248° − 180°), namely, $68\frac{1}{2}°$.

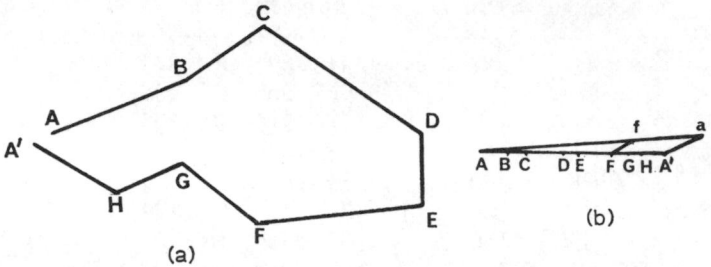

FIG. 24. (*a*) *Plot of traverse to scale.* (*b*) *Distribution of closing error.*

Referring to Fig. 24, which is a copy of an accurate plot by scale and compass:

(*vii*) The closing error is AA', which scales $45\frac{1}{2}$ paces, in the direction N 65° E.

(*viii*) The error at any other point may be obtained by plotting points A, B, C, D, E, F, G, H, and A' to scale along a base-line so that $AB = 300$ paces, $BC = 200$ paces, etc., and $AA' = 1950$ paces. The closing error $A'a$ is then drawn to another convenient scale in the direction N 65° E. The error at F is Ff drawn parallel to $A'a$ and scaled off as 33 paces. Hence,

Correct point F is 33 paces N 65° E of its plotted position.

It should also be noted that plotting by scale and protractor is not very accurate, and the slightest carelessness will give vastly different closing errors.

THE STANDARD TRAVERSE TABLE

Calculations should be arranged (*a*) so that they are easy to check, and (*b*) so that the probability of making mistakes is reduced to a minimum. Frequently, both objects are achieved by presenting the calculations in an orderly, generally accepted, tabular form. The full standard traverse table, the recognised method of presenting traverse calculations, is given in Specimen Question 16; the column headings must be learned, even though it is not always necessary to reproduce all the columns in answers to examination questions.

The purpose of the traverse table is to derive the co-ordinates of each station point so that the traverse can be plotted like a graph, a much more accurate method than plotting by protractor and scale.

As a preliminary to Specimen Question 16, it is necessary to define the reduced bearing (usually abbreviated to R.B.) of a line. This is the angle, never greater than 90°, between the line and the north or south directions, whichever is the nearer (*see* Fig. 25). The R.B. of any line such as I in the first quadrant will be N θ_1° E, and the R.B. of a line such as II in the second quadrant will be S θ° E. Similarly, the R.B.s of III and IV are respectively S θ_3° W and N θ_4° W.

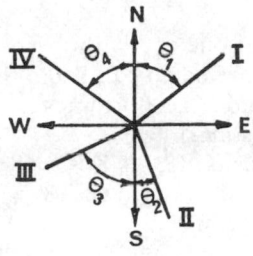

FIG. 25. *How reduced bearings are reckoned.*

SPECIMEN QUESTION 16

The mean observed internal angles and measured sides of a closed traverse *ABCDA* (in anti-clockwise order) are as follows:

Angle	Observed value	Side	Measured length (m)
DAB	97° 41′	AB	222.1
ABC	99° 53′	BC	583.4
BCD	72° 23′	CD	399.7
CDA	89° 59′	DA	521.0

Adjust the angles, compute the latitudes and departures assuming that *D* is due north of *A*, adjust the traverse by the Bowditch method, and give the co-ordinates of *B*, *C* and *D* relative to *A*.

SOLUTION

The calculations are performed in the following order, but are finally presented in the traverse table, Table 2.

(*a*) Column (1) gives the theodolite station and column (2) the station the instrument is to be at next, e.g. on the first line of Table 2 the instrument is at *A* and will be moved to station *B* next.

(*b*) The included angles, column (3), are summed, giving a total of 359° 56′. The number of sides, *N*, is 4 and therefore $(2N - 4) \times 90°$ equals 360° 00′. The closing error of 4 minutes is therefore distributed by adding 1 minute to each angle, the revised values appearing in column (4). This column is also summed as a check.

(*c*) The reduced bearings are calculated, starting with line *DA*. Since its W.C.B. is 180°, *A* is due south of *D*. The corrected included angle at A is $97°$ $42'$. Reference to Fig. 26 (a) shows that, for AB, the

$$\text{R.B.} = \text{S} (180° - 97°\ 42')\ \text{E} = 82°\ 18'\ \text{E}.$$

The corrected included angle at B is $99°$ $54'$. The reader should, by making a sketch and using the reduced bearing of AB, satisfy himself that for BC the reduced bearing is N $17°$ $36'$ E and that D is due west of C.

(d) For each line the co-ordinates of one end with respect to the other are calculated. The co-ordinates are taken parallel to the north and east directions and are called latitudes (i.e. northings and southings) and departures (eastings and westings).

As the length of $AB = 222.1$ m and its R.B. is S $82°$ $18'$ E, B is $222.1 \cos 82°\ 18' = 29.8$ m south of A, and $222.1 \sin 82°\ 18' = 220.5$ m east of A.

These calculations are performed by logarithms because this leads to a convenient way of setting out the calculation; in Table 2 a calculator would, however, be more accurate. The latitude is given the symbol $\triangle y$ and the departure $\triangle x$. Thus,

$$\log \triangle y = \log (\text{length}) + \log \cos (\text{R.B.})$$
and $$\log \triangle x = \log (\text{length}) + \log \sin (\text{R.B.})$$

The logarithmic operations are recorded, in a compact manner, in column (8) and the results, the consecutive co-ordinates, are entered in columns (9), (10), (11) and (12).

(e) Since the traverse is closed, the sums of columns (9) and (10) should be equal, as should the sums of columns (11) and (12). A closing error is almost inevitable. In this example the sum of column (9) exceeds that of column (10) by 5.4. One method of eliminating this error is Bowditch's method, in which the error is distributed in proportion to the length of the sides.

For example, on line AB the southing is increased by $5.4 \times 222.1/1726.2 = 0.7$ m; and on line BC the northing is decreased by $5.4 \times 583.4/1726.2 = 1.8$ m. Notice that the sum of the adjustments:

$$0.7 + 1.8 + 1.3 + 1.6 = 5.4, \text{ the closing error}$$

This agreement is absolutely necessary, and sometimes may be achieved only after a further correction of perhaps 0.1 m in one of the adjustments. Notice also that these adjustments are calculated to only the same absolute accuracy as the measured lengths, the nearest 0.1 m this example.

The closing error in the departures is distributed in exactly the same way; e.g. on line AB the easting is increased by $2.8 \times 222.1/1726.2 = 0.3$ m.

Another method of distributing the closing error is by the transit

TABLE 2. TRAVERSE TABLE FOR SPECIMEN QUESTION 16

(1) From	(2) To	(3) Included angle	(4) Corrected included angle	(5) W.C.B.	(6) Reduced bearing θ	(7) Length (l)	(8) log Δy log cos θ log l log sin θ log Δx	(9) N	(10) S	(11) E	(12) W	(13) N	(14) S	(15) E	(16) W	(17) Lat. (N)	(18) Dep.(E)
									Consecutive co-ordinates				Corrected consecutive co-ordinates			Independent co-ordinates	
A	B	97° 41′	97° 42′		S 82° 18′ E	222.1	1.4737		29.8	220.5			30.5	220.8		−30.5	220.8
							1.1251 2.3456 1.9951									(B)	
							2.3427		(+0.7)	(+0.3)							
B	C	99° 53′	99° 54′		N 17° 36′ E	583.4	2.742	556.2		176.4		554.4		177.3		523.9	398.1
							1.979 2.766 1.480									(C)	
							2.246	(−1.8)		(+0.9)							
C	D	72° 23′	72° 24′		West	399.7			0.0		399.7		1.3		399.0	522.6	−0.9
									(+1.5)		(−0.7)					(D)	
D	A	89° 59′	90° 00′	180°	South	521.0			521.0	0.0			522.6	0.9		0	0
									(+1.6)	(+0.5)						(A)	
Totals:		359° 56′	360° 00′			1726.2		556.2	550.8 Diff. = 5.4	396.9	399.7 Diff. = 2.8	554.4	554.4	399.0	399.0		

rule, which is described in, for example, Wilson's *Land Surveying*.

(*f*) By adding or subtracting the adjustments to the consecutive co-ordinates, the corrected consecutive co-ordinates are obtained: columns (13), (14), (15) and (16). Each of these columns is summed for a check.

(*g*) Taking station A (any other station will do equally well) as the origin with co-ordinates (0, 0), the independent co-ordinates of B are:

$$\text{latitude} = -30.5; \text{departure} = 220.8$$

Since the corrected consecutive co-ordinates of C from B are 554.4 N and 177.3 E, the independent co-ordinates of C are:

$$\text{Latitude} = -30.5 + 554.4 = 523.9 \text{ m}$$
and $$\text{Departure} = 220.8 + 177.3 = 398.1 \text{ m.}$$

The independent co-ordinates of D are obtained by adding -1.3 and -399.0 to these totals.

Notice that the independent co-ordinates of A are given in the table on the same line as "To" A, and the independent co-ordinates of B are on the same line as "To" B, etc.

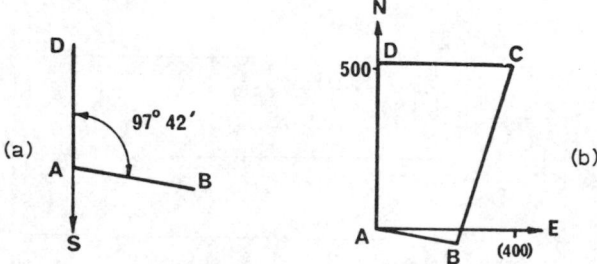

FIG. 26. (*a*) *Calculating the reduced bearing of* AB. (*b*) *Traverse and the co-ordinate axes.*

(*h*) A useful exercise for the reader is to plot the traverse from the independent co-ordinates to give a shape similar to that in Fig. 26 (*b*), which is drawn approximately to scale.

ASSESSMENT OF ACCURACY

Closing error in latitudes = 5.4 m and in departures = 2.8 m.
∴ linear closing error = $\sqrt{(5.4^2 + 2.8^2)} = 6.0$ m.

Since total length of sides = 1726.2, accuracy = $6/1726 \simeq 1/300$, which is reasonable for chain surveying.

In theodolite work a relatively large error of 1 minute is equivalent to a linear error of only $\pi/(180 \times 60) = 1/3500$, approximately.

Hence, assuming the traverse is by theodolite and chain, the closing error is reasonable, and would have been incurred almost entirely during chaining.

AREA ENCLOSED BY A TRAVERSE

The area enclosed by a traverse can always be split into a number of rectangles and right-angled triangles and then calculated. A more orderly procedure, however, is to use the following formula:

$$\text{Area} = \tfrac{1}{2}\sum(x_A + x_B)(y_B - y_A)$$

which means:

$$\text{Area} = \tfrac{1}{2}[(x_A + x_B)(y_B - y_A) + (x_B + x_C)(y_C - y_B) + (x_C + x_D)(y_D - y_C) + \ldots]$$

The right-hand side of this equation has one term for each side of the traverse, x_A, x_B, x_C, ... being the longitudes of the vertices A, B, C, ... and y_A, y_B, y_C, ... being their latitudes. The term $(x_A + x_B)$ is sometimes called the double longitude of the side AB.

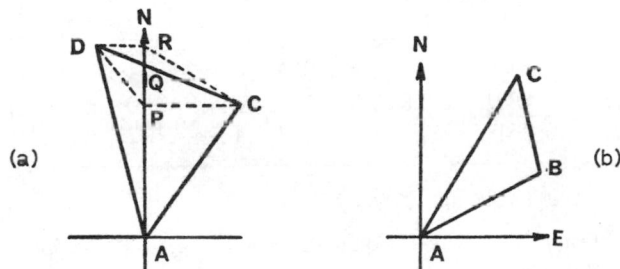

FIG. 27. *Two simple traverses, (a) and (b), for which, it can easily be demonstrated, the "double longitude" formula gives the correct area.*

The proof of the formula applied to triangle ACD in Fig. 27 (a), is:

$$\tfrac{1}{2}(x_A + x_C)(y_C - y_A) = \triangle ACP$$
$$\tfrac{1}{2}(x_C + x_D)(y_D - y_C) = \tfrac{1}{2}x_C(y_D - y_C) + \tfrac{1}{2}x_D(y_D - y_C)$$
$$= \triangle PCR - \triangle PRD$$
$$\tfrac{1}{2}(x_D + x_A)(y_A - y_D) = \triangle ARD$$

Summing these three results (and bearing in mind that $\triangle CRD = \triangle PRD$, both being on the same base and having the same height),

$$\tfrac{1}{2}\sum(x_A + x_B)(y_B - y_A) = \triangle ACP + \triangle PCR - \triangle PRD + \triangle ARD$$
$$- \text{area } ACRD - \triangle PRD$$
$$= ACRD - \triangle CRD$$
$$= \triangle ACD$$

It should be noticed that x_D and $(y_A - y_D)$ are both negative, A being the origin of the graph.

The proof that the formula holds for triangle ABC in Fig. 27 (b) is simpler and is left as an exercise for the reader. By adding results similar to these two the formula can be verified for any traverse.

In Specimen Question 17 the traverse table is modified slightly. Some columns are omitted because the simple way the data are presented makes this possible, whereas three columns are added for the purpose of calculating the area. Notice that the signs of the double longitudes and the latitudes of the sides are important. Notice also that as the given data are mainly of 3-figure accuracy, all calculations have been performed on a slide-rule. A calculator set to this degree of accuracy would also suffice.

SPECIMEN QUESTION 17

A plot of land is up for sale and there is some doubt about its area. As a quick check, a compass traverse is run along the boundaries. Determine the area enclosed by the traverse from the following data:

Line	Bearing	Metres	Line	Bearing	Metres
A–B	195°	528	E–F	343°	788
B–C	275°	548	F–G	5°	653
C–D	$182\frac{1}{2}°$	813	G–H	$80\frac{1}{2}°$	1421
D–E	$261\frac{1}{2}°$	1293	H–A	$102\frac{1}{2}°$	778

SOLUTION. *See* Table 3.

TABLE 3. TRAVERSE TABLE FOR SPECIMEN QUESTION 17

Line	Reduced bearing (θ)	Length (m)	$\cos \theta$ $\sin \theta$	N	S	E	W	Double longitude	Products +	Products −
AB	S 15° W	528	0.966 0.259		510		137	-137	70 000	
BC	N 85° W	548	0.087 0.996	48			546	-820		39 400
CD	S $2\frac{1}{2}°$ W	813	0.999 0.044		812		36	-1402	1 140 000	
DE	S $81\frac{1}{2}°$ W	1293	0.148 0.989		192		1280	-2718	522 000	
EF	N 17° W	788	0.956 0.292	754			230	-4228		3 185 000
FG	N 5° E	653	0.996 0.087	650		57		-4401		2 862 000
GH	N $80\frac{1}{2}°$ E	1421	0.165 0.986	235		1401		-2943		691 000
HA	S $77\frac{1}{2}°$ E	778	0.216 0.976		168	760		-782	131 000	
			Totals:	1687	1682	2218	2229		1 863 000	6 777 000

From Table 3, the sum of the "Products" is:

$$1\,863\,000 - 6\,777\,000 = -4\,914\,000 \text{ m}^2$$

Hence, $\frac{1}{2}\Sigma(x_A + x_B)(y_B - y_A) = \frac{1}{2} \times 4\,914\,000 = 2\,457\,000 \text{ m}^2$

∴, since 1 hectare (ha) = 10 000 m^2, Area = 245.7 ha

i.e. to three significant figures, Area = 246 ha

NOTES

(*i*) The calculations for the first eight columns are performed in the same order as in Specimen Question 16.

(*ii*) As a check on the arithmetic the four columns for the consecutive co-ordinates have been summed, but the closing errors have not been distributed. This omission is permissible, as it does not impair the accuracy of the final answer.

(*iii*) Double longitudes are calculated as follows:

$$x_A + x_B = 0 - 137 = -137 \ m$$

for *A* is taken as the origin and the difference in longitude is that *B* is 137 m west of *A*.

Similarly, the longitude of *C* is $-137 - 546$, and therefore,

$$x_B + x_C = -137 - 137 - 546 = -820 \ m$$

Similarly,

$$x_C + x_D = -820 - 546 - 36 = -1402$$
$$x_D + x_E = -1402 - 36 - 1280 = -2718$$
$$x_E + x_F = -2718 - 1280 - 230 = -4228$$
$$x_F + x_G = -4228 - 230 + 57 = -4401 \text{ etc.}$$

(*iv*) The "Products" are calculated by multiplying the double longitude by the latitude of the line, taking into account whether it is +(N) or −(S).

e.g. for *DE*,   product $= -2718 \times (-192) = +522\,000$

SINES OF SMALL ANGLES (*see* also p. 122, Chapter 8)

In order to complete a traverse table it is sometimes necessary to find, for example, log sin 0° 43′, which is not given in common 4-figure log sine tables. The procedure is to use the property that, for small angles, sin $\theta = \theta$ radians which, in 4-figure tables, is true for values of θ up to about 4° (although in 7-figure tables it is true only for angles < 30′). The same technique can be used with calculators.

Using 4-figure tables, sin 0° 43′ = 43/60 × π/180 radians.

hence   log sin 0° 43′ = log (43/60 × π/180) = $\overline{2}$.0971

A similar difficulty is encountered in evaluating log cos θ when θ is near 90°, but it can be overcome by using the same property, e.g.:

log cos 89° 47′ 23″ = log sin (90° − 89° 47′ 23″) = log sin 0° 12′ 37″

but   12′ 37″ = 757 seconds = 757/3600 × π/180 radians.

∴ log cos 89° 47′ 23″ = log sin 757″

$$= \log (757/3600 \times \pi/180) = \bar{3}.5646$$

EXAMINATION QUESTIONS

1. The field results for a closed traverse are:

Line	W.C.B.	Length (m)	Line	W.C.B.	Length (m)
AB	0° 00′	83	DE	160° 55′	101
BC	63° 49′	123	EF	264° 02′	67.5
CD	89° 13′	110	FA	258° 18′	199.5

The observed values of the included angles check satisfactorily, but there is a mistake in the length of a line. Which length is wrong and by how much?

[*Hint.* Find the closing error graphically. Its direction is roughly parallel to the line of incorrect length.]

2. The traverse table below refers to a closed traverse run from station D through O, G and H and closing on D. The whole circle bearing of O from D is 06° 26′ and G and H lie to the west of the line O D.

Compute the latitudes and departures of O, G and H with reference to D as origin, making any adjustments necessary.

Observed internal angles		Length in Metres	
HDO	79° 47′	DO	547.7
DOG	102° 10′	OG	939.8
OGH	41° 11′	GH	840.2
GHD	136° 56′	HD	426.5

3. The co-ordinates of the boundary stations A, B, C, D and E of a five sided piece of land are given below. It is required to divide the plot into two pieces of equal area by a line from station C to a point P on the line AE. Compute the distance AP and the area of each plot in hectares.

Station	A	B	C	D	E
X (northings) (m)	0	150	350	230	0
Y (eastings) (m)	0	− 50	80	240	200

[*Hint*. Drop a perpendicular CN to meet AE in N, find the area each side of CN, and construct a triangle CNP with an area equal to half the difference of these areas.]

4. A closed traverse, $ABCDE$, measured clockwise, is surveyed with a compass and chain, the angles being measured to the nearest 10 minutes and the lengths of the lines to the nearest 0.1 m. The following readings are obtained:

Line	Whole circle bearing	Length (m)
AB	28° 00′	300.8
BC	324° 30′	165.5
CD	40° 50′	302.6
DE	159° 30′	587.2
EA	265° 10′	482.2

Draw up a form of traverse sheet, determine the latitudes and departures of the lines, and balance the traverse.

5. The table below gives the lengths and uncorrected Relative Latitudes and Departures of the lines of a traverse $ABCDA$.

Line	Length (m)	Relative latitude (m)	Relative departure (m)
AB	162.24	+ 104.00	+ 124.52
BC	246.00	− 188.22	+ 158.38
CD	151.87	+ 147.55	+ 35.96
DA	324.78	− 62.80	− 318.65

Sketch the traverse and use the Bowditch method to correct the Relative Latitudes and Departures and calculate the co-ordinates of B, C and D relative to A as origin.

6. The table below gives the uncorrected relative latitudes and departures of a traverse $ABCDEFA$. The closing error is suspected to be due almost entirely to an error in one of the angles of the traverse.

Line	Relative latitude (m)	Relative departure (m)
AB	+303.00	+110.40
BC	+55.75	+320.10
CD	-148.50	-102.00
DE	-153.66	+181.30
EF	-267.02	-44.40
FA	+207.44	-465.07

Neglecting error from any other cause:

(a) determine, in metres, the closing error in the traverse;
(b) determine which internal angle is most likely to be erroneous;
(c) calculate the error in this angle.

7. A traverse is run between two points, P and Q, the co-ordinates of which are 1684.27 m (E) 2123.15 m (N) and 2089.26 m (E), 2225.13 m (N) respectively. The following observations are recorded.

Line	Length (m)	Bearing
PA	263.4	36° 15′
AB	196.1	132° 41′
BQ	103.8	75° 27′

Compute the traverse, adjust the closing error using Bowditch's method and hence obtain the adjusted co-ordinates of A and B.

8. A traverse is carried out using a 20 m chain and a theodolite which measures to 30 seconds. The traverse is run between points P and Q the co-ordinates of which are 2122.1 m (E), 1247.4 m (N) and 2423.0 m (E), 1562.8 m (N) respectively. The following results are recorded:

Side	Length	Bearing
PK	216.2	17° 24′
KQ	247.4	68° 36′

When computed, it is discovered that a mistake has been made. Compute the traverse and suggest where and how the mistake has occurred. Correct the mistake, adjust any remaining error and hence determine the co-ordinates of point K.

9. The co-ordinates of corners of a triangular plot of land ABC are as follows:

Point	Easting (m)	Northing (m)
A	1110	1226
B	1468	1612
C	1752	1340

The area is to be divided into two equal parts by a line XY, where X is a point on AB and 25 m from A and Y is a point on the side BC. Calculate the co-ordinates of point Y.

CHAPTER 5

THE USE OF THE THEODOLITE

II. IN TACHEOMETRY AND TRIANGULATION

FIXED HAIR TACHEOMETRY

The following pages deal with tacheometry and the use of the theodolite as a tacheometer. Tacheometry is employed chiefly as a fast method of contouring, for, by reading both vertical and horizontal verniers in addition to the stadia and cross hair readings on a levelling staff, it is possible to calculate both the staff position and the reduced level of the foot of the staff.

A suitable procedure is indicated in Fig. 28 and described below. $A, B, C, \ldots$ are successive instrument stations.

(a) The theodolite is set up at A and, with the horizontal scale reading $0°$, is pointed to B. Staff readings are taken first at B then at

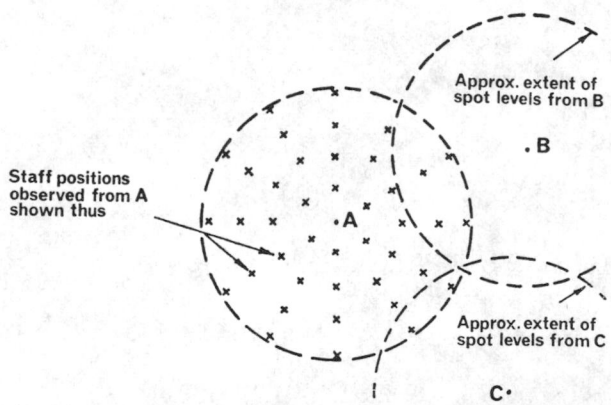

FIG. 28. *Tacheometry procedure.*

fairly regular intervals (perhaps every 10 paces) along the line BA, the vertical angle being adjusted only when necessary.

(b) After freeing the upper clamp the telescope is turned until the horizontal scale reads, say, $30°$. The upper clamp is then fixed and another line of staff readings is taken, again at regular intervals of distance.

(c) This procedure is continued, taking staff readings along lines at, say, 60°, 90°, 120°, ... 330° from AB so that spot levels can be calculated for points covering a circle, centre A. Probably, check readings are also taken on a staff at C.

(d) The tacheometer is then moved to B, C, ... in turn. At each the routine is the same, starting with readings on a staff held at a neighbouring station, and leads to spot levels covering an approximate circle with its centre at the instrument station.

(e) It is necessary to take a bearing on one of the lines, BC, for example, so that the survey can be oriented.

(f) Usually at least one set of staff readings is taken on to a bench mark so that reduced levels can be related to Ordnance Datum.

It should be noted that, for every staff position, the following observations are recorded: instrument station, height of instrument, the outer stadia readings and the mid-reading, the horizontal scale reading, the vertical scale reading and, for key points, the amount by which the altitude bubble is off centre so that the vertical angle reading can be adjusted to its true value if this bubble is not central.

To survey the same area with a dumpy level would involve much more fieldwork, for: (a) more instrument stations would be required if the terrain were at all hilly, because the telescope cannot be used parallel to the ground as it can in tacheometry, and (b) the staff positions would have to be surveyed independently, by chain, for example.

THE PRINCIPLE OF TACHEOMETRY

Horizontal sights with staff normal to line of collimation

When the levelling staff is observed through the telescope the outer stadia give two distinct staff readings; the length of staff between these is the staff intercept, s. The distance between the stadia lines, i, is

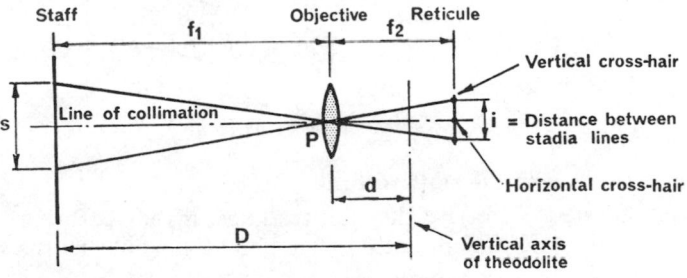

FIG. 29 *The symbols used in establishing the principle of tacheometry.*

usually 2 or 3 mm. Light rays passing through P, the principle point, are straight lines.

Consequently, in Fig. 29, $s/i = f_1/f_2$ by similar triangles.

A fundamental law of optics states that $1/f = 1/f_1 + 1/f_2$. Eliminating f_2 between these equations,

$$1/f = 1/f_1(s/i + 1)$$
$$\therefore f_1 = f(s/i + 1) = cs + f,$$

where $c = f/i$, and is called the multiplying constant.

$$\therefore D = f_1 + d = cs + (f + d)$$
or $$D = cs + k$$

where $k = (f + d)$, and is called the additive constant.

In this theory, f is the focal length of the objective and is therefore a constant for any particular instrument, as are i and d. Consequently, c and k are instrument constants, c being usually 100.

When the tacheometer is correctly set up the vertical axis is over the instrument station and, therefore, D is the distance between the instrument station and the staff, provided the telescope is horizontal and the staff is vertical.

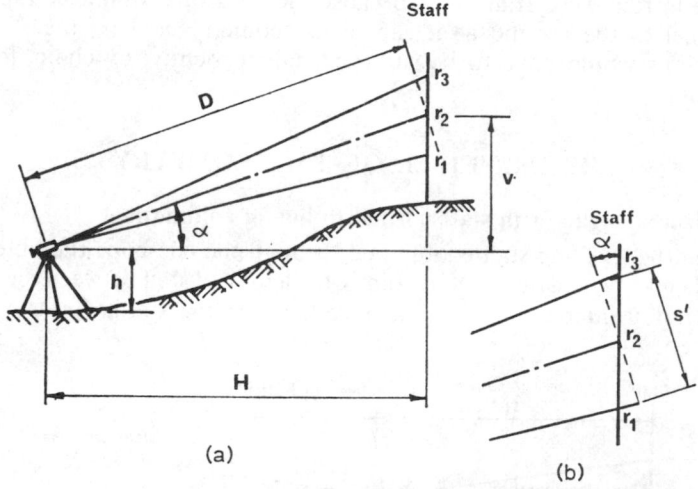

FIG. 30. *Inclined tacheometry sights with the staff vertical.*

Inclined sights with the staff vertical

In Fig. 30 (a) r_1, r_2, r_3 are the staff readings. If the staff were held normal to the line of collimation, s', Fig. 30 (b), would be the staff intercept.

$$\therefore D = cs' + k$$

Moreover, if the tiny error due to the divergence of the outer lines of sight is neglected,

$$s' = s \cos \alpha$$

where $s = r_3 - r_1 = $ actual staff intercept.

Also, horizontal distance from the instrument station to the staff is

$$H = D \cos \alpha$$
$$v = D \sin \alpha$$

If h is the height of instrument (i.e. the vertical distance between the station peg and the trunnion axis) the foot of the staff is a distance $V = v + h - r_2$ above the instrument station.

Combined formulae

The formulae given above are sufficient for solving most problems on tacheometry. Sometimes these formulae are combined to give:

$$H = cs \cos^2 \alpha + k \cos \alpha$$
$$v = cs \sin \alpha \cos \alpha + k \sin \alpha = \tfrac{1}{2}cs \sin 2\alpha + k \sin \alpha$$

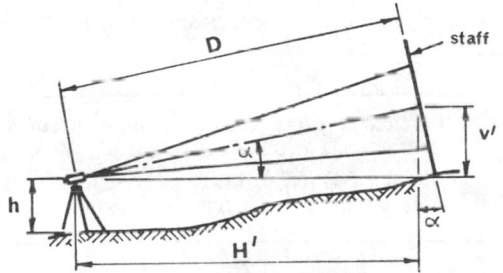

FIG. 31. *Inclined tacheometry sights with staff normal to line of sight.*

Inclined sights with staff normal to line of collimation

Figure 31 shows the main measurements. For rough work the distance between the instrument and staff is frequently taken as H' instead of $H' + r_2 \sin \alpha$, and the difference in height between the instrument station and the foot of the staff is frequently taken as $V = v' + h - r_2$ instead of $v' + h - r_2 \cos \alpha$. The errors so incurred are fairly small if α is less than about $4°$.

In this method errors will arise if the staff is inclined not exactly normal to the collimation, while in the staff vertical method errors will arise if the staff is not exactly vertical. The staff normal method has the advantage that, for a given angular displacement of the staff, such errors are likely to be smaller than in the staff vertical method. Against this, much less is required of the staffman if the staff is used vertically and, for this reason, the method is usually preferred.

THE MULTIPLYING AND ADDITIVE CONSTANTS

Determining the constants

The values of c and k are usually supplied with the instrument by the manufacturers. Should it be necessary to determine them (for example, on an old tacheometer), the simplest procedure is to set up the theodolite on a flat, approximately horizontal field and take readings on a vertical staff held in turn at two positions, each a known distance from the instrument. The constants can then be evaluated by solving simultaneous equations in c and k as illustrated in Specimen Question 18. The method becomes self-checking if, in addition, readings are taken on a staff held at a third position which is also at a known distance from the instrument—*see* question 2, p. 85.

Many tacheometers are fitted with an anallatic (a word frequently spelt as "anallactic") lens, a device which reduces k to zero, so that then, $D = cs$.

SPECIMEN QUESTION 18
A tacheometer is to be used to obtain the difference in level between two points A and B. The instrument is set up at I and the following data recorded:

Point	Vertical angle	Readings on vertical staff		
A	$-6°\ 24'$	3.605	2.920	2.235
B	$-8°\ 30'$	1.975	1.095	0.215

In order to obtain the constants of the tacheometer, the instrument is used afterwards with the line of collimation horizontal and sighting on to a levelling staff held vertically at 25 and 50 m from the instrument. The outer readings recorded are 1.700, 1.210 and 2.420, 1.430 respectively. If the level of A is assumed 100.00, determine the level of B and the horizontal distance of A from the instrument.

SOLUTION
The tacheometry constants are determined from $D = cs + k$:

when $D = 25$ m, $s = 1.700 - 1.210 = 0.490$ m, and $\therefore\ 25 = 0.49c + k$.
when $D = 50$ m, $s = 2.420 - 1.430 = 0.990$ m, and $\therefore\ 50 = 0.99c + k$

hence
$$25 = 0.50c$$
$$\therefore\ c = 50 \text{ and } k = 0.5 \text{ m}$$

Line IA: In Fig. 32, $\alpha = 6°\ 24'$, and therefore,
$$D = 50(3.605 - 2.235)\cos 6°\ 24' + 0.5 \text{ m} = 68.58 \text{ m}$$

hence $\qquad v = 68.58 \sin 6° \ 24' = 7.59$ m

and, since mid-reading $= 2.920$ m

$\qquad A$ is $(7.59 + 2.920 - h) = (10.510 - h)$ m below I

Line IB: In the sketch, $\alpha = 8° \ 30'$,

$\qquad \therefore \ D = 50(1.975 - 0.215) \cos 8° \ 30' + 0.5$ m $= 87.54$ m

hence $\qquad v = 87.54 \sin 8° \ 30' = 12.86$ m

Mid-reading $= 1.095$ and $\therefore \ B$ is $(12.86 + 1.095 - h)$
$$= (13.955 - h) \text{ m below } I$$

Hence B is $(13.955 - h) - (10.510 - h) = 3.455$ m below A, and, since R.L. of $A = 100.00$ m.

$$\text{R.L. of } B = 100.00 - 3.455 = \underline{\underline{96.555 \text{ m}}}$$

Horizontal distance from I to A: For line IA, $D = 68.58$ m and $\alpha = 6° \ 24'$.

$$\therefore \ H = 68.58 \cos 6° \ 24' = \underline{\underline{68.16 \text{ m}}}$$

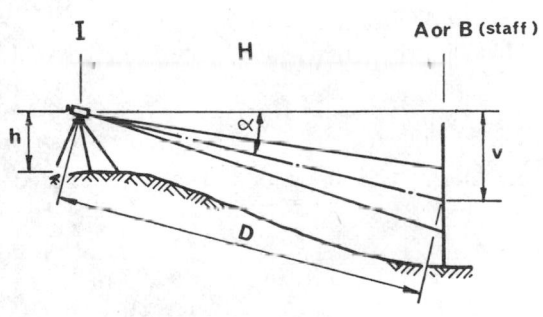

FIG. 32

SPECIMEN QUESTION 19

Readings taken with a tacheometer that has a multiplying constant of 100 and an additive constant of 0.5 m were recorded as follows:

Instrument at	Staff at	Vertical angle	Stadia readings	Remarks
P	Q	30° 00′ elevation	$\left\{ \begin{array}{c} 1.73 \\ 2.65 \\ 3.57 \end{array} \right\}$	Vertical staff

Although the calculations were made on the assumption that the staff was vertical, it was in fact held at right angles to the collimation. Compute the errors, caused by this mistake, in the calculation of horizontal and vertical distances from the instrument to the foot of the staff. Give the sign of each error.

If the collimation is not horizontal, is it preferable to have the staff vertical or at right-angles to the collimation? Give reasons for your preference.

SOLUTION

$$s = 3.57 - 1.73 = 1.84$$

With staff vertical,

$$D = 1.84 \times 100 \times \cos 30° + 0.50 = 159.8 \text{ m}$$
$$\therefore H = 159.8 \cos 30° = 138.4 \text{ m}$$
and
$$v = 159.8 \sin 30° = 79.90 \text{ m}$$
$$\therefore V = h + 79.90 - 2.65 = h + 77.25 \text{ m}$$

With staff normal to collimation,

$$D' = 1.84 \times 100 + 0.50 = 184.50 \text{ m}$$
$$H' = 184.5 \cos 30° = 159.8 \text{ m}$$

$$\therefore \text{True horizontal distance} = 159.8 + 2.65 \sin 30° = 161.1 \text{ m}$$
$$\text{Deduct calculated value:} \qquad = 138.4 \text{ m}$$

$$\text{Error} = \underline{\underline{22.7 \text{ m}}}$$

and the correction is to be <u>added</u> to the calculated horizontal distance.

Also,
$$v' = 184.5 \times \sin 30° = 92.25 \text{ m}$$
$$\therefore V' = h + 92.25 - 2.65 \cos 30°$$

hence   True vertical distance $= h + 92.25 - 2.30 = h + 89.95 \text{ m}$
$$\text{Deduct calculated value:} \qquad h + 77.25 \text{ m}$$

$$\text{Error} = \underline{\underline{12.70 \text{ m}}}$$

and the correction is to be <u>added</u> to the calculated vertical distance.

The descriptive part is answered on p. 71.

TACHEOMETRY SHEET

Tacheometry fieldwork as described on page 68 is usually recorded in a table called a tacheometry sheet, an abridged version of which is given in Specimen Question 20. For contouring work two more columns are necessary, one giving the bearing of the line of sight and the other giving the horizontal distance from instrument to staff. The positions of the columns in the table can be interchanged to some extent.

TABLE 4. TACHEOMETRY SHEET FOR SPECIMEN QUESTION 20

(1)	(2)	(3)	(4)	(5)	(6)	(7)
Station	Height of instrument (h)	Staff position	Outer stadia		Inclination (α)	Mid-reading (m)
A	1.35	B	1.50	0.65	$+0°\ 54'$	1.08
B	1.48	C	1.81	1.08	$-2°\ 54'$	1.45
C	1.46	D	1.57	0.69	$+2°\ 48'$	1.13
D	1.40	A	1.85	0.97	$-1°\ 48'$	1.41

	(8)	(9) Gen. number (cs)	(10) $\frac{1}{2}cs \sin 2\alpha$	(11)	(12)	(13) R.L. at foot of staff
Station	h − m			Rise	Fall	
A	0.27	85	+1.34	1.61		173.61
B	0.03	73	−3.69		3.66	169.95
C	0.33	88	+4.29	4.62		174.57
D	−0.01	88	−2.76		2.77	171.80

Notes on the tacheometry sheet, Table 4
(a) The table is designed for an instrument with an anallatic lens so that the formulae: $H = cs \cos^2 \alpha$; $v = \frac{1}{2}cs \sin 2\alpha$ apply, the symbols being as already defined. The quantity cs is called the generating number.

(b) The rises and falls are calculated on a completely different basis from those in the "rise and fall" method of levelling, and give the difference in reduced level between the instrument station and the foot of the staff.

(c) The figures in columns (8) and (10) may be positive or negative. On each line the two figures are added together, taking account of the signs, the answer, if positive, being a rise and, if negative, a fall.

SPECIMEN QUESTION 20

The following readings were taken with an anallactic tacheometer set up at each station in turn and a staff held vertically on the forward station, the forward station from D being A.

Station	Height of instrument	Stadia readings			Inclination (elevation +ve)
A	1.35	1.50	1.08	0.65	+0° 54′
B	1.48	1.81	1.45	1.08	−2° 54′
C	1.46	1.57	1.13	0.69	+2° 48′
D	1.40	1.85	1.41	0.97	−1° 48′

The reduced level of A is 172.0 m and the constant of the tacheometer is 100.

Determine the reduced levels of B, C and D, adjusted to close on A, indicating and justifying your method of adjustment.

SOLUTION

The calculation for this problem is presented in Table 4. As an example of the order in which the figures are derived, consider line AB

Generating No. $= cs = 100(1.50 - 0.65) = 85$ [column (9)]
$\therefore \frac{1}{2}cs \sin 2\alpha = \frac{1}{2} \times 85 \sin 1° 48' = 1.34$ [column (10)]

$h = 1.35$ [column (2)] and $m = 1.08$ [column (7)], these quantities being given in the question.

$\therefore h - m = 1.35 - 1.08 = 0.27$[column (8)]
          Rise $= 1.34 + 0.27 = 1.61$
$\therefore$ R.L. of $B = $ R.L. of $A + 1.61 = 173.61$ m   [column (13)]

The calculations for the other four lines are similar, and lead ultimately to the value 171.80 m for the R.L. of A, instead of the given value 172.00 m.

Hence closing error $= 172.00 - 171.80 = 0.20$ m.

$\therefore$ To adjust the R.L.s, add $0.20/4 = 0.05$ m to each:

          R.L. at $B = 173.61 + 0.05 = 173.66$ m
          R.L. at $C = 169.95 + 2 \times 0.05 = 170.05$ m
          R.L. at $D = 174.57 + 3 \times 0.05 = 174.72$ m

The closing error is distributed equally between the four points for the following reasons:

(a) Staff readings are read to 0.01 m independently of distance.

(*b*) An error in the vertical angle gives an error proportional to length in the R.L.

(*c*) A slight error in the multiplying constant gives an error proportional to length in the R.L.

Since the generating numbers are roughly the same for each line, the lengths of the four lines are roughly the same, and therefore the errors may be expected to be roughly the same for each.

THE BEAMAN STADIA ARC

After working through Specimen Question 20 it will be evident that the field observations of an extensive tacheometric survey lead to a great deal of arithmetic in the office. Tacheometry tables, instead of the commoner 4-figure trigonometrical tables, can be used to speed up the calculations. Better, a tacheometer incorporating one of the many devices for automatically reducing observations can be employed. The most commonly examined of these devices is the Beaman stadia arc, the principle of which is indicated in Fig. 33. The diagram shows a Beaman stadia arc for a tacheometer with a multiplying constant of 100. The telescope must have an anallatic lens.

FIG. 33. *Beaman stadia arc.*

The central scale is the vertical protractor which turns with the telescope, and which is found on all theodolites. In Fig. 33 the vertical scale reads $0°$ when the telescope is horizontal.

The H-scale gives values of $c(1 - \cos^2\alpha)$ against the corresponding value of α, the angle of inclination.

The V-scale gives values of $\frac{1}{2}c \sin 2\alpha$ against the corresponding value of α.

Thus, the horizontal distance from instrument station to staff is

$$(100 - \text{H-scale reading}) \times s$$

and vertical distance from telescope to the mid-reading on the staff is

$$\text{V-scale reading} \times s$$

where s = staff intercept.

The scales are easily constructed from the values in Table 5, suitably extended. On some instruments, however, the scales are calibrated differently from those shown in Fig. 33.

TABLE 5. CONSTRUCTING THE BEAMAN STADIA ARC

$100(1 - cos^2 \alpha)$	$cos^2 \alpha$	$cos \alpha$	α	$50\ sin\ 2\alpha$	$sin\ 2\alpha$	2α	α
1	0.99	0.995	$5\frac{3}{4}°$	0	0	$0°$	$0°$
5	0.95	0.9747	$13°$	5	0.100	$5\frac{3}{4}°$	$2\frac{7}{8}°$
10	0.90	0.9487	$18\frac{1}{2}°$	10	0.200	$11\frac{1}{2}°$	$5\frac{3}{4}°$

SPECIMEN QUESTION 21

What is the principle of, and the field procedure for, the Beaman stadia arc, as fitted to a telescopic alidade for use in plane tabling?

A telescopic alidade so fitted was aligned on to a graduated staff held vertically on a summit. The tangent screw was adjusted until the vertical scale of the arc read 25 elevation; then the horizontal scale read 6.7 and the staff readings were 2.10, 1.55 and 1.00 m. Determine to the nearest 0.1 m the difference in level and the horizontal distance between the summit and the plane table.

SOLUTION

Plane tabling is a method of surveying in which the plan is drawn point by point as the survey proceeds. The basic equipment consists of the plane table, i.e. a drawing board mounted on a tripod with some form of levelling head, and the alidade, which consists essentially of a pair of sights mounted on and parallel to a straight-edge. In this question the alidade is telescopic and is fitted with a Beaman stadia arc so that the following procedure is not suitable for the simpler type of alidade.

Assuming part of the survey has been accomplished from previous stations, the instrument is moved to the new station and set up there so that the (previously) plotted position of the station is vertically above the station peg. In addition, the plane table is levelled and correctly oriented.

The staffman holds the staff at significant points on the survey. With the straight-edge of the alidade against the plotted station point, the surveyor then sights the staff at each point in turn, making a note of the stadia readings and, if levels are required, the mid-reading. By applying the principles described on p. 57 he determines the distance to the staff from the staff intercept and the H-scale reading, and can therefore plot the staff position immediately, for the direction of the staff is given by the straight-edge. If levels are required he determines these from the V-scale reading.

Since the H-scale reading = 6.7, and staff intercept = 2.10 − 1.00
                                                     = 1.10 m,
Horizontal distance = (100 − 6.7) × 1.10 = 102.6 m

Difference in level between telescope and mid-reading is:

V-scale × 1.10 = 27.5 m
Deduct mid-reading: =   1.55 m

Required difference in level = 25.95 m
                             = 26.0 m to the nearest 0.1 m

Note. The multiplying constant is assumed to be 100, the usual value.

OTHER METHODS OF TACHEOMETRY

The great majority of tacheometrical surveys are conducted by the methods described in the foregoing pages. To get these methods into perspective, however, it is necessary to appreciate that there are two systems of tacheometry: the stadia system and the tangential system. The stadia system can again be subdivided into two categories which are called fixed-hair and movable-hair tacheometry respectively— and of these only fixed-hair tacheometry has so far been dealt with in this book.

In movable-hair tacheometry, as the name implies, the distance between the stadia lines, i, can be increased or decreased so that, obviously, a special tacheometer which has this facility must be used. The formula given on p. 70 still applies: i.e. $D = (f/i)s + k$. In this method, however, i is the variable on the right-hand side of the equation and the staff intercept, s, is fixed, sights usually being taken on to a special staff with two fixed targets, a distance s apart.

The fundamental characteristic of the tangential system is that it requires two pointings of the telescope for each computation of distance or height. In Fig. 30(a) for example, the staff reading r_2 would be observed against the central cross hair and the angle of elevation, α, would be observed and recorded; no staff readings would be taken against the outer stadia. Instead the angle of elevation would be increased to β, say, and a staff reading such as r_3 again taken against the central cross hair. From the two pointings the dimensions H and v can be calculated as follows:

$$\frac{v}{H} = \tan \alpha$$

and, now, with $r_3 - r_2 = s$,

$$\frac{v + s}{H} = \tan \beta$$

Eliminating H between these two equations:

$$\frac{v}{\tan \alpha} = \frac{v + s}{\tan \beta}$$

$$\therefore \quad v \frac{\tan \beta - \tan \alpha}{\tan \alpha \tan \beta} = \frac{s}{\tan \beta}$$

$$\therefore \quad v = \frac{s.\tan \alpha}{\tan \beta - \tan \alpha}$$

and, since $H = v \cot \alpha$,

$$H = \frac{s}{\tan \beta - \tan \alpha}$$

TRIANGULATION

The highest degree of accuracy in surveying is obtainable by the method of triangulation, in which the three angles of a triangle are measured by theodolite. By arranging the station points to form a network of triangles, vast areas can be surveyed and, provided that one base line is measured (forming a side of one of the triangles), sufficient information is obtained for plotting the stations. In the case of a national survey, topographical detail is obtained by local surveys based on the triangulation (or trig) points.

Closing error

In triangulation, as in other kinds of survey, a closing error will be inevitable, and must be distributed among the measured angles. This adjustment is governed by certain conditions which, for the case of triangles forming a polygon (i.e. the most general case), are as follows.

(*a*) The sum of the angles of each triangle must be 180° (+ spherical excess, a correction which must be applied if the area of the triangle exceeds about 180 km²).

(*b*) The sum of the angles at the "centre" must be 360°.

(*c*) $\Sigma \log \sin l = \Sigma \log \sin r$ (*see* Fig. 34).

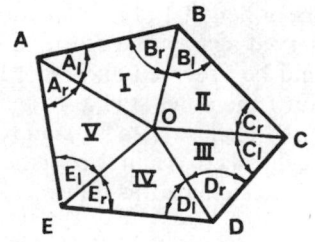

FIG. 34. $\Sigma \log \sin l = \Sigma \log \sin r$.

Proof that $\Sigma \log \sin l = \Sigma \log \sin r$

So that there are no ambiguities it is necessary that, given AB, the length of CD should have the same value whether it is calculated by means of triangles I, II and III or, alternatively, by means of triangles I, V, IV and III (Fig. 34).

By the sine formula:

(clockwise) $\dfrac{BO}{AB} = \dfrac{\sin A_l}{\sin \angle BOA}$ (anti-clockwise) $\dfrac{AO}{AB} = \dfrac{\sin B_r}{\sin \angle BOA}$

$\dfrac{CO}{BO} = \dfrac{\sin B_l}{\sin C_r}$ $\dfrac{EO}{AO} = \dfrac{\sin A_r}{\sin E_l}$

$\dfrac{CD}{CO} = \dfrac{\sin \angle DOC}{\sin D_r}$ $\dfrac{DO}{EO} = \dfrac{\sin E_r}{\sin D_l}$

$$\dfrac{CD}{DO} = \dfrac{\sin \angle DOC}{\sin C_l}$$

$$\dfrac{BO}{AB} \times \dfrac{CO}{BO} \times \dfrac{CD}{CO} = \dfrac{CD}{AB} = \dfrac{\sin A_l \,.\, \sin B_l \,.\, \sin \angle DOC}{\sin \angle BOA \,.\, \sin C_r \,.\, \sin D_r}, \text{ and}$$

$$\dfrac{AO}{AB} \times \dfrac{EO}{AO} \times \dfrac{DO}{EO} \times \dfrac{CD}{DO} = \dfrac{CD}{AB} = \dfrac{\sin B_r \,.\, \sin A_r \,.\, \sin E_r \,.\, \sin \angle DOC}{\sin \angle BOA \,.\, \sin E_l \,.\, \sin D_l \,.\, \sin C_l}$$

CD/AB can have only one value, and therefore the right-hand sides of these two equations must be the same, i.e.:

$$\dfrac{\sin A_l \,.\, \sin B_l}{\sin C_r \,.\, \sin D_r} = \dfrac{\sin B_r \,.\, \sin A_r \,.\, \sin E_r}{\sin E_l \,.\, \sin D_l \,.\, \sin C_l}$$

Cross-multiplying and taking logs of both sides:

$\log \sin A_l + \log \sin B_l + \log \sin C_l + \log \sin D_l + \log \sin E_l$
$\quad = \log \sin A_r + \log \sin B_r + \log \sin C_r + \log \sin D_r + \log \sin E_r$

which is usually written:

$$\sum \log \sin l = \sum \log \sin r$$

where l and r denote left-hand and right-hand angles, respectively and the Greek letter, Σ, denotes "sum of".

It should be noted that when the triangles do not form a polygon only conditions 1 and 3 are applied, e.g. in Question 9 at the end of this chapter.

The adjustments are made by the method of equal shifts, illustrated in Specimen Question 22, in which:

(a) a closing error in any triangle is distributed equally between the three angles of the triangle;

(b) a closing error at the centre is distributed equally between all the angles at the centre;

(c) as (b) upsets the adjustment in (a) a second small adjustment has to be made to two angles in each triangle; and

(d) any difference between $\Sigma \log \sin l$ and $\Sigma \log \sin r$ is distributed in such a way that all angles concerned are altered by the same amount, except that, if the correction to the left-hand angles is positive, the correction to the right-hand angles will be negative, and vice versa.

SPECIMEN QUESTION 22

In order to demonstrate how a triangulation is adjusted by the method of equal shifts, consider a figure which consists of a triangle ABC with a central (internal) point D and in which the following fictitious angles are given as "observed angles".

$$BAD = ABD = CBD = BCD = ACD = 30° \ 00'$$
$$ADB = BDC = CDA = 120° \ 00'$$
$$CAD = 33° \ 00'$$

Although the error in $\triangle ADC$ is so large that a gross mistake appears to have been made, adjust the angles of triangulation (to the nearest minute) to give a consistent figure.

What are the five equations of conditions to which the adjusted angles must conform?

SOLUTION

Referring to Fig. 35, the conditions to be satisfied are:

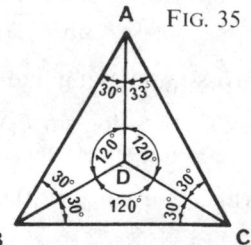

Fig. 35

(a) $\angle ABD + \angle BDA + \angle DAB = 180°$
(b) $\angle BCD + \angle CDB + \angle DBC = 180°$
(c) $\angle CAD + \angle ADC + \angle DCA = 180°$
(d) $\angle BDA + \angle CDB + \angle ADC = 360°$
(e) $\Sigma \log \sin l = \log \sin r$, where

$\qquad \angle ABD, \ \angle BCD, \ \angle CAD$ are l, the left-hand angles
and $\qquad \angle DAB, \ \angle DBC, \ \angle DCA$ are r, the right-hand angles

First adjustment. Deduct 1° from each angle in $\triangle CAD$, making their sum 180°. This reduces $\angle ADC$ to 119° and the sum of the angles at the centre to 359°; hence:

Second adjustment. Add 20 minutes to each of $\angle BDA$, $\angle CDB$, $\angle ADC$. This makes the sum of the angles at the centre 360°, but alters the sum of the angles of each triangle to the value of 180° 20′; hence:

Third adjustment. Deduct $20/2 = 10$ minutes from each left-hand and each right-hand angle.

Conditions (a)–(d) are now satisfied. The adjusted values of the angles are entered in column (2) of Table 6. The other columns of Table 6 present a tabular method by means of which the fourth adjustment makes $\Sigma \log \sin l = \Sigma \log \sin r$.

$$\Sigma \log \sin l - \Sigma \log \sin r = \overline{1}.1156 - \overline{1}.0767 = 0.0389 = 389 \text{ units.}$$

Let the left-hand angles be decreased by x minutes each and right-hand angles increased by x minutes each.

Then, $\log \sin \angle ABD$ is decreased by $2.2x$, $\log \sin \angle DAB$ is increased by $2.2x$ and so on. Hence, to make the totals of columns (3) and (4) balance,

$$x(2.2 + 2.2 + 2.0) + x(2.2 + 2.2 + 2.4) = 389$$

i.e. $\qquad 13.2x = 389 \text{ units}$

and $\qquad x = 389/13.2 = 29.4 \text{ minutes}$

Hence, the final values, column (6), are obtained by deducting 29 minutes (to the nearest minute) from left-hand angles and adding 29 minutes to right-hand angles in column (2).

TABLE 6. TRIANGULATION ADJUSTMENT
(see Specimen Question 22)

(1)	*(2)*	*(3)*	*(4)*	*(5)*	*(6)*
	Value after first				
Angle	*3 adjustments*	*log sin l*	*log sin r*	*Diff. l'*	*Final value*
ABD	$29°\ 50'$	$\overline{1}.6967$		2.2	$29°\ 21'$
BDA	$120°\ 20'$				$120°\ 20'$
DAB	$29°\ 50'$		$\overline{1}.6967$	2.2	$30°\ 19'$
BCD	$29°\ 50'$	$\overline{1}.6967$		2.2	$29°\ 21'$
CDB	$120°\ 20'$				$120°\ 20'$
DBC	$20°\ 50'$		$\overline{1}.6967$	2.2	$30°\ 19'$
CAD	$31°\ 50'$	$\overline{1}.7222$		2.0	$31°\ 21'$
ADC	$119°\ 20'$				$119°\ 20'$
DCA	$28°\ 50'$		$\overline{1}.6833$	2.4	$29°\ 19'$
	Totals:	$\overline{1}.1156$	$\overline{1}.0767$		

NOTE
Columns (3), (4) and (5) are obtained from 4-figure log sine tables. For angles between $29°$ and $30°$ the "mean difference" for 5 minutes

is 11, so that the value entered in column (5) is $11/5 = 2.2$. For more accurate work 7-figure log sine tables are used and column (5) is replaced by "Diff. 1 sec."

Braced quadrilateral

Figure 34 is sometimes referred to as a *centre point polygon*.

Another common case where the method of equal shifts is applied is shown in Fig. 36 and is referred to as a braced quadrilateral.

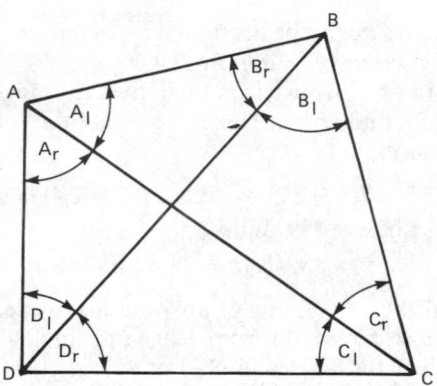

FIG. 36. *A braced quadrilateral.*

For plane figures the adjustment of the angles is governed by the following conditions:

- (a) $A_r + A_l + B_r + B_l + C_r + C_l + D_r + D_l = 360°$;
- (b) $A_r + D_l = B_l + C_r$, and
  $A_l + B_r = C_l + D_r$;
- (c) $\Sigma \log \sin l = \Sigma \log \sin r$.

EXAMINATION QUESTIONS

1. With the aid of diagrams and the notation given below, derive from first principles the formula for calculating by tacheometry the horizontal distance H and difference in height at ground level V between two stations A and B. B is higher than A and the angle of elevation is $\theta°$. The staff is to be held vertically at B.

f = focal length,
c = distance from vertical axis to principal point,
k = stadia constant,
s = staff intercept,
I = height of instrument above ground level,
h = height of axis reading on staff.

2. The readings given below were made with a tacheometer to determine its constants, the telescope being horizontal for each reading. Find the value of the constants from these observations.

Distance of staff from instrument	Stadia readings (m)
100 m	1.123, 1.616, 2.110
150 m	1.778, 2.519, 3.260
200 m	1.390, 2.379, 3.367

Describe briefly a method by which values of horizontal distance and the difference in height between two stations can be obtained automatically without the use of tables.

3. The observations recorded below were made with a tacheometer having an anallatic lens. Compute the gradient of the slope from A to B.

Ins. station	Observing to	Stadia readings on vertical staff			Vertical angle to centre reading	Bearing from north
		Upper	Centre	Lower		
O	A	2.97	1.71	0.45	$-8°\ 20'$	$228°$
	B	2.93	1.90	0.87	$10°\ 40'$	$48°$

At the instrument station the ground level was 278.40 m and the height of the instrument axis was 1.5 m.

4. Readings are taken, using a tacheometer with a multiplying constant of 100 and an additive constant zero, first on a staff held vertically on a bench mark 210.63 m above datum and then on point P. The following observations are recorded:

Station sighted	Stadia readings	Centre hair	Reduced vertical angle
Bench Mark	2.680	1.400	$+04°\ 24'$
	0.120		
Point P	2.005	1.055	$-03°\ 12'$
	0.105		

The positive and negative signs in the vertical angle column refer to angles of elevation and depression respectively.

Calculate, deriving from first principles, any formulae used:

(a) The reduced level of P.
(b) The horizontal distance from the instrument to P.

5. A tacheometer having a multiplying constant of 100 and an additive constant of zero is set up over a station A and the following readings are recorded.

Inst. at	Staff at stn.	Horizontal circle reading	Vertical circle reading	Stadia readings (m)		
A	B	0° 00′	0° 00′	0.800	1.093	1.385
A	C	42° 15′	+7° 00′	1.444	1.810	2.176

If the reduced level of station A is 54.000 m A.O.D. and the instrument axis is 1.52 m above A, calculate:

(a) the horizontal distance between B and C;
(b) the reduced level of B and C.

The staff was held vertically for all readings.

6. The following tacheometric observations were taken with a theodolite having an anallactic telescope with a constant of 100. The staff was vertical.

Inst. at	Reading on stn.	Height of inst. axis above stn. (m)	Stadia readings (m)			Angle of elevation
F	G	1.48	1.650	1.300	0.950	+5° 00′
G	H	1.55	1.145	0.825	0.505	−8° 00′

Determine:

(a) the horizontal distance between F and G and between G and H;
(b) the reduced levels of G and H if that of F was 100.000 m.

7. Points X, Y and Z are situated on the summits of three adjacent hills. A tacheometer is set up over point X and readings are taken on to a staff held vertically on stations Y and Z in turn. The tacheometer has a multiplying constant of 100 and no additive constant. The following observations are recorded:

Instrument station	Staff station	Stadia readings	Centre hair reading	Vertical circle	Horizontal circle
X	Y	2.885 0.260	1.572	$+3°\ 15'$	$11°\ 42'$
X	Z	2.905 0.515	1.710	$-2°\ 45'$	$73°\ 18'$

The positive and negative signs in the vertical circle column indicate angles of elevation and depression respectively.

Calculate the horizontal distances XY, YZ and XZ and also the difference in height between Y and Z.

8. A braced quadrilateral forms parts of a triangulation network. Explain clearly the angle and side conditions which must be taken into account when adjusting the observations taken in this figure.

Draw up a table for, and hence explain how the observations are adjusted by, the method of equal shifts.

9. In the triangulation network shown in Fig. 37 all the angles have been observed and the sides DH and GC measured as base and check base respectively, with the following results:

Triangle DHO	Triangle HGO	Triangle GCO
$D\ =79°\ 47'\ 05''$	$H = 77°\ 28'\ 58''$	$G\ =82°\ 22'\ 17''$
$H\ = 58°\ 32'\ 35''$	$G = 36°\ 02'\ 38''$	$C\ = 71°\ 29'\ 47''$
$O\ = 41°\ 40'\ 05''$	$O = 66°\ 28'\ 48''$	$O\ = 26°\ 08'\ 17''$
$DH = 426.58$ m		$GC = 486.83$ m

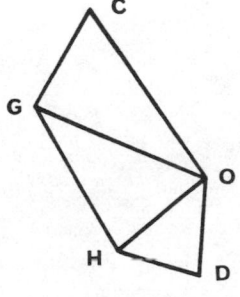

FIG. 37

Adjust the observed angles by "equal shifts" to give a consistent figure.

[*Hint.* The conditions to be satisfied are:

(a) Sum of the angles of each triangle = $180°$.
(b) Log DH + log sin $\angle ODH$ + log sin $\angle OHG$ + log sin $\angle COG$
= log GC + log sin $\angle HOD$ + log sin $\angle HGO$
+ log in $\angle GCO$.]

CHAPTER 6

APPLICATIONS OF SIMPLE GEOMETRY

I. USE OF THE PROPERTIES OF THE CIRCLE

USEFUL THEOREMS

A number of surveying problems can be treated as problems in geometry, the theorems given in Table 7 being particularly useful. For the sake of brevity, the proofs have not been given, but the reader is strongly recommended to work through these, referring if necessary to any standard school geometry textbook.

SPECIMEN QUESTION 23

It is required to join two straights, XY and YZ, with a simple circular curve, but there is an obstruction between the two straights which makes it imperative that the curve should not extend beyond a line PQ joining the straights. Determine the tangent distance XY and the radius of a curve that will be tangential to the three straights XY (on which the point P lies), YZ (on which the point Q lies) and PQ.

The angle of intersection $XYZ = 135°$ and the angles made by the straight PQ with XY and YZ are $XPQ = 150°$ and $PQZ = 165°$. The distance $YP = 100$ m.

SOLUTION

The configuration is shown in Fig. 38. From the tangent properties of a circle (*see* Table 7, 3)

$$\angle POT = \tfrac{1}{2}\angle TPY = 15°,$$

and
$$\angle QOT = \tfrac{1}{2}\angle TQY = 7\tfrac{1}{2}°.$$

Hence, in $\triangle QTO$, $\quad QT = R \tan 7\tfrac{1}{2}° = 0.1317R$,

and, in $\triangle PTO$, $\quad PT = R \tan 15° = 0.2679R$.

$$\therefore PQ = PT + TQ = 0.3996R.$$

Applying the sine formula to $\triangle PYQ$,

$$\frac{PY}{\sin 15°} = \frac{PQ}{\sin 45°}$$

and, since $PY = 100$ m (given),

$$PQ = 100 \times \frac{\sin 45°}{\sin 15°} = 0.3996R$$

$$\therefore R = 683.6 \text{ m}$$

and
$$XY = R \tan 22.5° = 283.1 \text{ m}$$

TABLE 7. A SUMMARY OF SOME USEFUL GEOMETRY THEOREMS

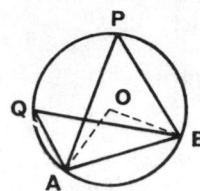

1. If AB is a chord and P, Q lie anywhere on the circle on the same side of AB, then

$$\angle APB = \angle AQB = \tfrac{1}{2}\angle AOB,$$

O being the centre of the circle.

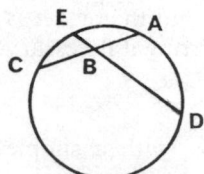

2. If AC and DE are any two chords of a circle which intersect at B, then

$$AB.BC = DB.BE$$

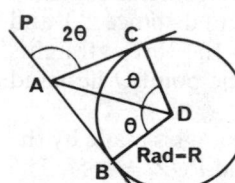

3. If AB and AC are tangents and D is the centre, then

$$AB = AC$$
$$\angle PAC = \angle BDC$$
$$\angle ABD = \angle ACD = 90°$$
$$\triangle ABD \equiv \triangle DCA$$

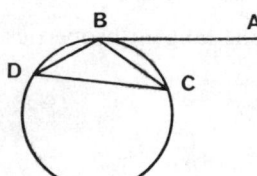

4. If AB is a tangent, BC is any chord, and D is any point on the circumference, in the alternate segment, then

$$\angle ABC = \angle CDB$$

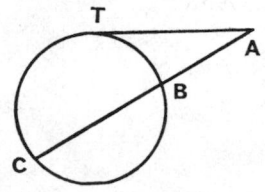

5. If AT is a tangent and ABC a secant then

$$AT^2 = AB.AC$$

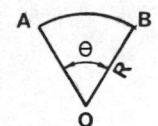

6. If arc AB subtends an angle of θ degrees at the centre, O, of a circle, radius R, then

$$\text{arc length} = \pi R \times \theta/180$$

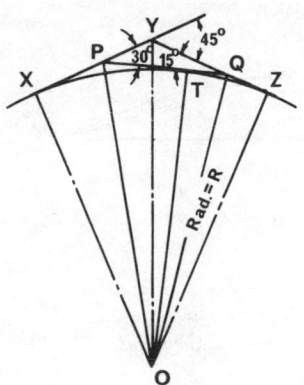

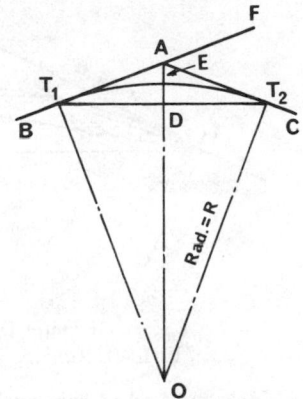

FIG. 38. *The configuration required in Specimen Question 23.*

FIG. 39. *The main geometrical features of a simple circular curve.*

SIMPLE CIRCULAR CURVES

Specification and setting out

Circular curves are used in roads and railways. Fig. 39 defines the main features associated with a circular curve. For this purpose the curve is shown connecting two straights BT_1 and T_2C, but it should be observed that circular curves can also join transition curves or other circular curves of a different radius. In Fig. 39 BA and AC are the tangents. A is the intersection point. T_1, T_2, are the tangent points. T_1T_2 is the long chord and DE is the versed sine of the curve. T_1ET_2 is the circular curve and is, in fact, an arc of a circle centre O and radius R. $\angle FAC$ is the deflection (or intersection) angle.

Circular curves can be specified in three ways:

(a) by the radius;
(b) by the angle subtended at the centre by a chord of standard length, often 30 m; and
(c) by the versed sine.

It is left as an exercise for the reader to prove that a 5° curve (based on a 30-m chord) is, in fact, a curve of about 344 m radius.

Fig. 40 shows one method of setting out circular curves, namely Rankine's method of tangential angles (sometimes called, ambiguously, "deflection angles"), in which a theodolite and chain, or tape, are used. Other methods of setting out, using chain and tape only or, alternatively, two theodolites, exist, but only the first is described as the two theodolite method is mainly of academic interest.

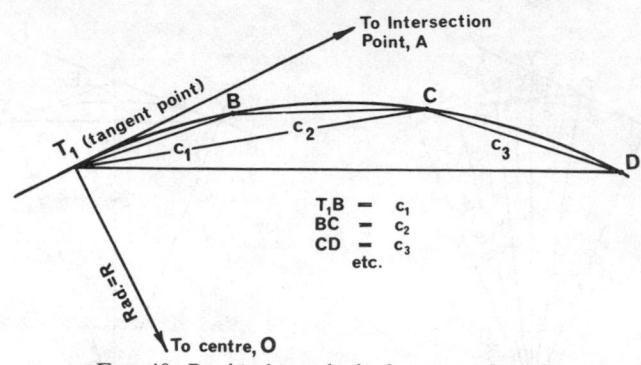

FIG. 40. *Rankine's method of tangential angles.*

Rankine's method of tangential angles

Suppose the chord lengths chosen for setting out the curve are $c_1, c_2, c_3, \ldots$, which for 4-figure accuracy should not greatly exceed $R/15$, then the deflection angles are:

$$\delta_1 = \angle AT_1B = 1718.9 \; c_1/R$$
and
$$\delta_2 = \angle BT_1C = 1718.9 \; c_2/R$$
and
$$\delta_3 = \angle CT_1D = 1718.9 \; c_3/R, \text{ etc. } (\textit{see} \text{ Fig. 40}).$$

Procedure

(a) Set up a theodolite at T_1 and, with the horizontal scale reading $0°$, point the telescope to A.

(b) Turn the telescope until the horizontal scale reading is δ_1 and, in this direction, chain a distance c_1 to give B, a point on the curve.

(c) Turn the telescope until the horizontal scale reading is $(\delta_1 + \delta_2)$ and chain a distance c_2 from B to intersect the line of sight at C, which will then be another point on the curve.

(d) Repeat the previous step but with scale readings $(\delta_1 + \delta_2 + \delta_3)$, $(\delta_1 + \delta_2 + \delta_3 + \delta_4)$, etc., until the distant tangent point, T_2, is established.

Proof of formulae

(a) Providing the chords are small compared with the radius, the approximation

$$\text{chord} = \text{arc length}$$

is permissible. Hence,

$$c_1 = \text{arc } T_1B = \pi R \times 2\delta_1/(60 \times 180), \text{ if } \delta_1 \text{ is in minutes}$$
i.e. $\delta_1 = 60 \times 90/\pi \times c_1/R = 1718.9 \; c_1/R \text{ minutes}.$

(b) Similarly, $\delta_2 = 1718.9 \; c_2/R \text{ minutes}$
and $\delta_3 = 1718.9 \; c_3/R \text{ minutes, etc.}$

Chord lengths

When setting out the centre-line of a proposed road or railway, pegs will be established at every chain length (every 30 m, say) and, provided the radius is large enough to justify their use, chords of 1 chain in length will be used to mark out a proposed curve. If the chainage of T_1 is, for example, 100 chains + 17 m (i.e. 3017 m) it may be convenient to make the first chord only 13 m so that the chainage of B will be 3030 m—although subsequently 30-m chords are to be used making the chainages of $C, D, \ldots$ 3060 m, 3090 m ... respectively. In these circumstances, the chainages are said to be carried through the tangent point T_1. The alternative is to make the first chord length the same as the following chords, of course.

The length of the last chord is determined by the length of the curve, which in turn depends on the deflection angle and radius.

Alternative Method—by chain and tape

A simpler method of setting out a circular curve, but at the cost of some accuracy, is to use chain and tape. The procedure, referring to the sketch in Fig. 40, is as follows.

(*i*) Choose a suitable radius and suitable chord lengths, c_1, c_2, c_3, etc.

(*ii*) Mark out the tangent from T_1 towards A by ranging it from the straight leading up to T_1.

(*iii*) Chain along the tangent and place an arrow at a distance of c_1 from T_1; from this arrow offset a distance of $o_1 = c_1^2/(2R)$ to give B, a point on the curve. (Remember an offset is always measured at right-angles to the main chainage.)

(*iv*) By ranging, extend T_1B. Chain along T_1B produced and place an arrow at a distance of c_2 from B; from this arrow offset a distance $o_2 = c_2(c_1 + c_2)/(2R)$ to give C, a point on the curve.

(*v*) By ranging, extend BC. Chain along BC produced and place an arrow at a distance of c_3 from C; from this arrow offset a distance $o_3 = c_3(c_2 + c_3)/(2R)$ to give D, a point on the curve.

(*vi*) Continue this routine until the second tangent point is fixed.

If "through" chainage is used the first two offsets will almost certainly be odd lengths but all the remaining offsets, excepting the last, will be equal in length. This follows because in practice c_1 will be short but $c_2 = c_3 = c_4 = \ldots = c$, the standard chord length, and the offset formulae become:

$$o_1 = c_1^2/(2R)$$
$$o_2 = c(c_1 + c)/(2R)$$
$$o_3 = o_4 = \ldots o = c^2/R$$

Proof that $o_1 = c_1^2/(2R)$

Let $\angle AT_1B = \alpha$, then the angle subtended by T_1B at the centre $= 2\alpha$

$$\therefore c_1 \simeq \text{arc length } T_1 B = R.2\alpha$$

i.e. $$\alpha = c_1/(2R)$$

Similarly, since $\angle AT_1B = \alpha$, $o_1/c_1 = \sin \alpha \simeq \alpha$

$$\therefore o_1 = c_1\alpha = c_1{}^2/2R$$

The proof of the other offset formulae is left as an exercise for the reader.

SPECIMEN QUESTION 24
The centre-line of a proposed railway consists of two straights joined by a curve of 600-m radius. The angle of deflection between the two straights is $26°$, and the chainage (increasing from left to right) of their intersection is 7367 m.

Calculate the deflection angles to the nearest $20''$ from the tangent for setting out the circular curve from the first tangent point by pegs at every 30 m chainage and check on to the second tangent point.

SOLUTION
See Fig. 41.

From theorem 3, Table 7,

$$AB = R \tan 13°$$
$$= 600.\tan 13° = 138.5 \text{ m}$$

Hence,

chainage at $A = 7367 - 138.5 = 7228.5 \text{ m}$

Since $\angle DBC = 26°$,

$$\text{arc } AC = \pi \times 600 \times \frac{26}{180} \qquad\qquad = \quad 272.3$$

$$\therefore \text{ chainage at } C \qquad\qquad\qquad = \overline{7500.8 \text{ m}}$$

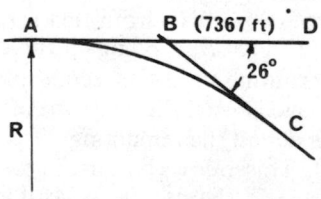

FIG. 41

The question requires that the chainage be "through" and therefore the first peg after A will be at the next convenient whole multiple of 30 m, namely 7260 m (because 7230 m would leave a chord of 1.5 m which would be impracticably short).

$$\therefore \text{ first chord} \qquad = 7260 - 7228.5 = 31.5 \text{ m}$$

Similarly, as the chainage of C is 7500.8 m, the last chord will be of 30.8 m in length but all other chords will be 30 m long.

For the first chord:

$$\text{tangential angle} = 1718.9 \times \frac{31.5}{600} = 90.22 \text{ minutes.}$$

For 30-m chords:

$$\text{increase in tangential angle} = 1718.9 \times \frac{30}{600} = 85.94 \text{ minutes.}$$

For the last chord:

$$\text{increase in tangential angle} = 1718.9 \times \frac{30.8}{600} = 88.22 \text{ minutes.}$$

Hence, the setting-out data are as shown in Table 8.

TABLE 8. SETTING OUT DATA FOR A CIRCULAR CURVE
(*see Specimen Question 24*)

(1) Chainage	(2) Increase in deflection angle	(3) Deflection angle	(4) Deflection angle to nearest 20"
(A) 7228.5 m	0	0° 00'	0° 00'
7260 „	90.22 mins	1° 30'.22	1° 30' 20"
7290 „	85.94 „	2° 56'.16	2° 56' 00"
7320 „	85.94 „	4° 22'.10	4° 22' 00"
7350 „	85.94 „	5° 48'.04	5° 48' 00"
7380 „	85.94 „	7° 13'.98	7° 14' 00"
7410 „	85.94 „	8° 39'.92	8° 40' 00"
7440 „	85.94 „	10° 05'.86	10° 06' 00"
7470 „	85.94 „	11° 31'.80	11° 31' 40"
(C) 7500.8 „	88.22 „	13° 00'.02	13° 00' 00"

CHECK

The deflection angle to C should be $26°/2 = 13°\ 00'$, so there is an error in the calculations of 00.02 minutes, which is negligible for 4-figure work.

NOTES

All the arithmetic is performed by 4-figure logs.

The figures in column (4) are running totals of the figures in column

(2); e.g. $4° \ 22'.10 = 90.22 + 2 \times 85.94$. This addition is often performed mentally and column (2) omitted.

CORRECTIONS FOR CURVATURE AND REFRACTION IN LEVELLING

For most purposes it is sufficient to take the Earth as being spherical, so that lines of constant altitude are really circles with their centres at the centre of the Earth. In levelling, because of this, a curvature error creeps in when the staff is at a greater distance than is usual from the instrument (e.g. the error is about 5 mm when the staff is at a distance of about 270 m).

In Fig. 42 the arc IB is a line of constant altitude passing through I, the telescope of the level. If it is assumed straight the line of collimation would be IA, and therefore AB would be the error in the staff reading due to curvature. As the Earth's atmosphere becomes less dense with increasing height, refraction causes the line of sight to curve back towards the Earth along the dotted line Ia. Ba is therefore the error due to curvature and refraction, and on average (it varies with climatic conditions and altitude, etc.) it is $6/7 \times$ value for curvature alone.

Referring to Fig. 42,
$$R = \text{radius of Earth in km}$$
By 5, Table 7,
$$IA^2 = AB(AB + 2R)$$
$$\simeq AB.2R$$
∴ taking $2R = 12\,740$ km, $AB = IA^2/12\,740$ km
Hence, if distance (IA) of staff from instrument is D km,

$$\text{Correction for curvature} = 1000D^2/12\,740 = 78D^2 \text{ mm}$$

Correction for curvature and refraction $= 6/7 \times 78D^2 = 67/D^2$ mm

It should be noted that the coefficient of refraction, k, is defined as

$$k = \frac{\angle AIa}{\theta},$$

see Fig. 42; θ is the angle at the centre of the earth subtended by IB. Then

$$aB = D(\tfrac{1}{2} - k)\theta = \frac{D^2}{2R}(1 - 2k)$$

k is usually taken as 0.072 but can vary by $\pm 10\%$ and, exceptionally, 100%.

DIP OF THE HORIZON

In Fig. 42 if IB now represents the surface of the sea, then I would be the horizon of an observer at A (if no allowance is made for refraction). The dip of the horizon is $90° - \angle BAI$.

The effect of refraction is to extend the observer's horizon and to decrease the dip slightly.

SPECIMEN QUESTION 25
In extending a triangulation survey of the mainland to a distant offlying island, observations were made between two trigonometric stations, the one 1000 m and the other 4000 m above sea level. If the ray from one station to the other just grazed the sea, what was the approximate distance between the stations: (a) neglecting the correction for refraction, and (b) allowing for it? Assume a spherical earth of diameter = 12 770 km.

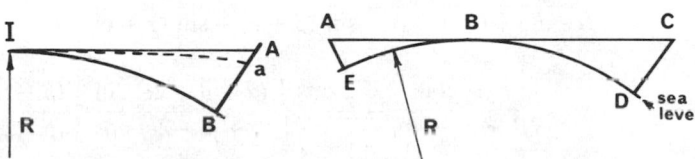

FIG. 42. *Error due to curvature and refraction.*   FIG. 43.

SOLUTION
Fig. 43 shows the configuration.
$AE = 1000$ m $= 1$ km
∴ by theorem 5, Table 7, $AB^2 = (1 + 12\,770) = 12\,771$
∴ $AB = 113.0$ km
Similarly, $BC^2 = 4 \times (4 + 12\,770)$ and ∴ $BC = 226.0$ km

Hence, $AC = AB + BC = \underline{339.0 \text{ km}}$ (a)

As mentioned above, the effect of refraction is equivalent to increasing the observer's altitude to $7/6 \times$ his altitude with straight-line vision. The distance to his horizon is therefore increased to $\sqrt{7/6}$ × distance with straight-line vision.

Hence: $\sqrt{7/6} \times 339.0 = \underline{366.1 \text{ km}}$ (b)

TRIGONOMETRICAL LEVELLING

The difference in level between two points A and B can also be calculated from observations of the vertical angle to the distant station measured by a theodolite at A or B. The best practice is to take readings with theodolites at A and B, a process known as reciprocal trigonometrical levelling, *see* Fig. 44 in which α and β are the observed zenith angles at A and B respectively. Equally, the angles of elevation at A and B could be used.

It is assumed that the refraction angles are equal at A and B and have the value e.

$$\frac{OB}{OA} = \frac{R + h_B}{R + h_A} = \frac{\sin(\beta + e)}{\sin(\alpha + e)}$$

$$\therefore \quad \frac{R + h_B - (R + h_A)}{R + h_A} = \frac{\sin(\beta + e) - \sin(\alpha + e)}{\sin(\alpha + e)}$$

and

$$\frac{R + h_B - (h_A)}{R + h_B + (R + h_A)} = \frac{\sin(\beta + e) - \sin(\alpha + e)}{\sin(\beta + e) + \sin(\alpha + e)}$$

i.e.

$$\frac{h_B - h_A}{2R + h_A + h_B} = \frac{2 \cos \frac{1}{2}(\alpha + \beta + 2e) \sin \frac{1}{2}(\beta - \alpha)}{2 \sin \frac{1}{2}(\alpha + \beta + 2e) \cos \frac{1}{2}(\beta - \alpha)}$$

$$= \tan \tfrac{1}{2}(\beta - \alpha) \cot \tfrac{1}{2}(\alpha + \beta + 2e)$$

But, from the geometry of the figure,

$$\alpha + \beta + 2e = 180° + \theta$$

$$\therefore \quad \cot \tfrac{1}{2}(\alpha + \beta + 2e) = \cot\left(90 + \frac{\theta}{2}\right) = \tan\frac{\theta}{2}$$

$\therefore$ since $h_A + h_B$ is very small compared with $2R$

$$h_B - h_A = 2R \tan \tfrac{1}{2}(\beta - \alpha) \tan\frac{\theta}{2}$$

and, since θ is small, $\quad 2R \tan\dfrac{\theta}{2} \simeq R\theta = d$

where d is the distance between stations. Hence

$$\text{difference in level} = h_B - h_A \simeq d \tan\frac{\beta - \alpha}{2}$$

It should be noted that the above theory assumes the vertical angles are measured from the transit axis of the distant theodolite. If this is not the case in practice, corrections have to be made for the height of collimation and the height of the target above the station.

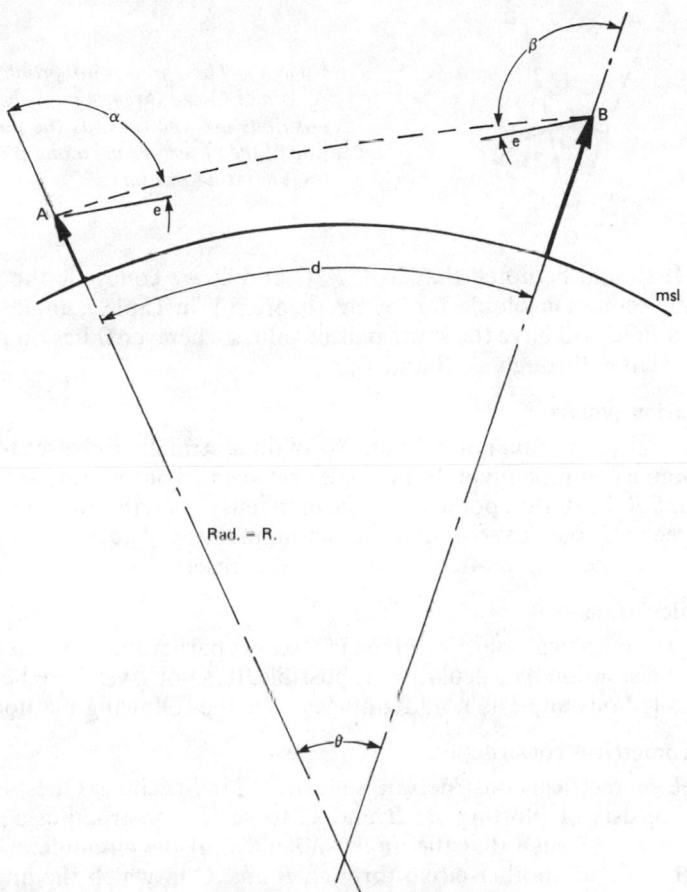

FIG. 44. *Reciprocal trigonometrical levelling.*

THE THREE-POINT PROBLEM

In an offshore underwater survey the position of the soundings (taken from a rowing boat, for example) can be located by measuring the angles subtended at the boat by AB and BC, A, B and C being three landmarks (*see* Fig. 45). The angles α and β are measured, one immediately after the other, by an observer in the boat using a sextant. Assuming he has a chart showing the positions of the three points, the observer can then determine his position by the following methods:

(*a*) by station pointer;
(*b*) by calculation;
(*c*) by a geometrical construction.

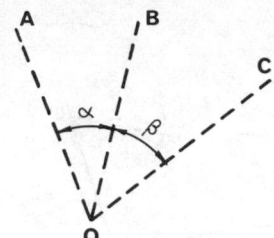

FIG. 45. *The three-point problem. A, B and C are three points whose positions are known. O is the position of the observer, and α and β are the angles he measures.*

It should be noted that if *A*, *B*, *C* and *O* are concyclic the problem becomes insoluble, for, by the theorem 1, in Table 7, angles *AOB* and *BOC* will have the same pair of values wherever *O* lies on an arc of a circle through *A*, *B* and *C*.

Station pointer

Basically, a station pointer consists of three straight arms free to turn about a common pivot. If the angles between adjacent arms are set to α and β the station pointer may be manoeuvred on the chart until the three arms pass over *A*, *B* and *C* simultaneously. The position of the pivot is then the plotted position of the observer.

Calculation

That a graphical solution (given in the next paragraphs) exist is proof that a solution by calculation is possible. It is not given here because it is tedious and has no advantages over the following method.

Geometrical construction

The geometrical construction is illustrated in Specimen Question 26. It consists of plotting *A*, *B* and *C* to scale, constructing a circle through *AB* such that the angle subtended at the circumference by *AB* is α and another circle through *B* and *C* in which the angle at the circumference is β. *O* is then the point of intersection of the two circles.

The process of solving the three-points problem is often named *resection*.

SPECIMEN QUESTION 26

The position of a launch when taking soundings offshore is located by resection from three fixed points *A*, *B* and *C*, which are clearly visible northwards from the launch. The co-ordinates in metres of the three fixed points are *A*, 1630 N and 40 E; *B*, 1610 N and 760 E; *C*, 1040 N and 1690 E.

The angle subtended at the launch by *A* and *B* is 24° 30′ and that by *B* and *C* is 41° 50′.

Plot *A*, *B* and *C* on squared paper to the scale of 20 000 to 1 and locate graphically the position of the launch, scaling its co-ordinates.

How would the subtended angles be measured and what device could be used aboard the launch to plot its position direct from their observed values?

SOLUTION

The points A, B and C are plotted to scale as indicated in Fig. 46. The reader should do this to the scale recommended in the question.

The position of the launch, L, is given by the intersection of two circles:

Join AB, draw AT_1 so that $\angle T_1AB = 24° \, 30'$ and draw BT_2 so that $\angle ABT_2 = 24° \, 30'$.

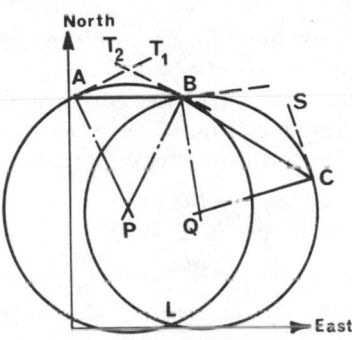

FIG. 46.

Draw lines through A and B, at right-angles to AT_1 and BT_2, to intersect at P. Then P is the centre and AP ($= BP$) the radius of one of the required circles, for,

$$\text{since} \qquad \angle T_1AB = \angle ABT_2 = 24° \, 30'$$
$$\angle BAP = \angle PBA = 90° - 24° \, 30'$$
$$\therefore \qquad \angle APB = 180° - 2(90° - 24° \, 30') = 2 \times 24° \, 30'$$
$$\text{but} \qquad \angle ABP = \text{the angle subtended by } AB \text{ at the centre,}$$

$\therefore$ the angle at the circumference,

$$\angle ALB = \tfrac{1}{2} \angle APB = 24° \, 30'$$

The second circle, centre Q and radius QC, is constructed in the same way, commencing, for example, by drawing SC at an angle of $41° \, 50'$ to CB. In this circle BC will subtend an angle of $41° \, 50'$ at the circumference.

Since the two circles intersect at only one point, other than B, this point, L, is the required point.

By scaling, the co-ordinates of L, in metres, are <u>10 N and 680 E.</u>

EXAMINATION QUESTIONS

1. (a) It is necessary to join two points A and B on the centre lines of two straight roadways by a circular curve. The angle between the two straights of the roadways at their intersection point X is 120° 00′. The circular curve has to have a radius of 150 m. The centre of the circle of which the curve is a part lies in the obtuse angle between the two straights. Calculate the:

  (i) tangent lengths AX and BX;
 (ii) length of the curve; and
(iii) angle subtended at the centre of the circule by the chord to the whole curve.

(b) Describe briefly how to set out the curve by the use of theodolite and tape.

2. The intersection point I of the two straight sections of road AB and CD, illustrated in Fig. 47, is inaccessible. The following observations are recorded from measurements taken at B and C.

$$\text{Angle } ABC = 133° \; 15′$$
$$\text{Angle } BCD = 107° \; 30′$$
$$BC = 417 \text{ m}$$

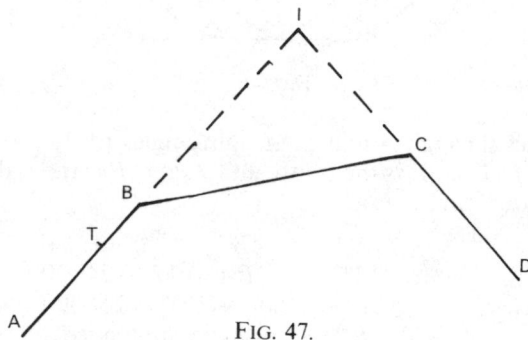

FIG. 47.

The two sections are to be connected by means of a circular curve which are to be tangential to AB at point T such that $TB = 39$ m. Calculate the radius of the required curve and also its total length.

3. The bearings of two straight sections of road AO and OB are 76° and 104° respectively. The sections are to be connected by a circular curve and two alternative schemes are being considered. The first scheme involves using point X as a tangent point where $OX = 212$ m. The second scheme involves the use of a 650 m radius curve. Calculate which scheme involves the shorter length of road and by how much.

4. It is required to set out a simple curve which will be tangential to three straights, two of which, XY and YZ, intersect at Y and the third runs from A on XY to B on YZ. The following angles are known: $AYB = 104°$ $36'$; $BAX = 148°$ $54'$; $ZBA = 135°$ $42'$. The chainage at A is 12 776 m and at Y is 14 296. What will be the chainage of the tangent point T on the straight XY and the through chainage of the point P where the curve touches the straight AB?

5. Figure 48 shows the centre lines of a reverse circular curve, comprising equal arcs AB and BC of equal radius, to be used to connect two parallel straight railway tracks. Find the radii of these two curves if the straight-line distance between the extreme tangent points is 760 metres.

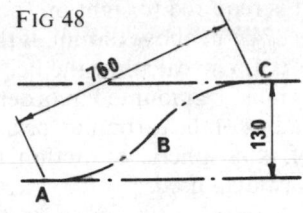

FIG 48

Calculate to the nearest 20 seconds the angles measured from the tangent for setting out from A to B with chord lengths of 50 m commencing at A.

6. In transferring levels across a river, pegs X and Y are placed on each bank 600 m apart.

Reciprocal observations are taken with the instrument near X and Y in turn and the following readings obtained, the bubble being central in each case.

Level near	Reading on: X	Y
X	2.070	2.220
Y	1.660	1.640

If the reduced level of X is 86.600 calculate:

(a) the reduced level of Y;
(b) the collimation error of the instrument, if any.

Take the diameter of the earth to be 12 740 km and refraction to be 1/7 curvature correction.

[*Hint*. The reciprocal levelling formula (*see* p. 45) still holds because the curvature error is the same in each direction.]

7. (a) Define the term coefficient of atmospheric refraction (k), and show that when $k = 0.07$ and the mean radius of the earth is 6370 km, the vertical displacement of a horizontal line of sight due to the combined effect of the earth's curvature and atmospheric refraction is given by

$$h = 0.068D^2,$$

where h = vertical displacement in metres and D = length of the line of sight in kilometres.

(b) A theodolite is set up at a station which is 272.06 m above datum. The height of the instrument above the ground is 1.40 m. A target (T) is sighted which is 8.5 km away, the angle of elevation of the telescope being recorded as $01°$ $15'$. The height of the target above the ground is 3.75 m. If the mean radius of the earth is taken as 6370 km and the coefficient of atmospheric refraction is 0.07, calculate the height of the ground at station T above datum.

8. A level is set up at a point P which is 363.4 m above datum. It is required to sight on to a station Q which is 12.4 km from P and is 374.7 m above datum. If the height of the level above the ground at P is 1.6 m, calculate the height above the ground at which the target Q must be mounted in order that it may be visible from P. The mean radius of the earth may be assumed to be 6375 km and the coefficient of atmospheric refraction 0.07. Derive, from first principles, any formulae used.

9. Explain the purpose of reciprocal observations when using a level.

Give a practical situation in which reciprocal observations would probably be used when levelling.

The levelling observations given below were made between two stations P and Q, 594.4 m apart:

Inst. at	Ht. of eye above stn. (m)	Reading On stn.	(m)
P	1.580	Q	0.043
Q	1.515	P	2.752

If the reduced level of P is 16.888 m determine

(i) the reduced level of Q
(ii) the collimation error of the level.

The error due to atmospheric refraction may be taken as $\frac{1}{7}$ of that due to earth curvature.

10. In a harbour survey A, B and C are three stations on shore and P is the location, in the harbour, of the end pile of a jetty under construction. Coordinates are as follows:

Stn.	x (East) (m)	y (North) (m)
A	0	0
B	470.00	105.50
C	750.25	500.60
P	150.00	620.10

P is to be set out by two theodolites located at A and B turning out the angles $B\hat{A}P$ and $A\hat{B}P$. It will then be checked by a theodolite at C, turning out the angle $B\hat{C}P$.

Calculate these three angles, to the nearest second, and also the horizontal distance of P from B.

11. Three markers, P, Q and R are erected in a straight line on a sea coast so that PQ measures 230.0 m, QR 125.0 m and PR 355.0 m. A boat offshore at a point X observed the angles PXQ and QXR as $38°\ 20'$ and $23°\ 40'$ respectively. Determine the distances of the boat from P and R.

In what circumstances would a fix be unobtainable using a method similar to this?

If the observed angles were such as to make a fix unobtainable, how could this be detected by the observer in the boat?

12. The known co-ordinates of three survey stations are as follows:

Station	Co-ordinates (m) East	North
A	49	389
B	288	389
C	487	274

From a new station P southwards from A, B and C, angles APB and BPC are observed to be $40°\ 40'$ and $34°\ 20'$ respectively.

Plot stations A, B and C to a scale of 1000 to 1 and then establish the position of P by a graphical method that does not involve trial and error.

CHAPTER 7

APPLICATIONS OF SIMPLE GEOMETRY

II. VERTICAL CURVES AND TRANSITION CURVES

When a new first-class road (or railway) is being planned, vertical curves and transition curves are introduced to make it safer and therefore faster. Parabolas and cubic parabolas, respectively, are almost always used for these curves. Both have well-known simple equations, and consequently the design of vertical or transition curves can often largely be treated as a problem in co-ordinate geometry. The properties and equations given in this chapter should of course be learned, as they reduce the amount of calculation involved.

VERTICAL CURVES

The equation of the parabola in its simplest form can be written

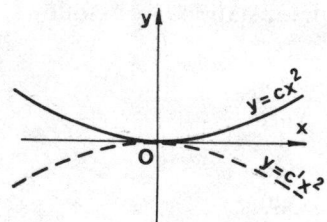

$$y = cx^2, \text{ where } c = \text{constant}$$

but this equation holds only if the axes are as shown in Fig. 49. Both axes must pass through the vertex O and the y axis must be the axis of symmetry of the parabola.

FIG. 49. *Two parabolas with the equation* y = Constant x².

Differentiating the above equation:

$$dy/dx = 2cx \propto x,$$

and therefore the slope of the parabola is proportional to x.

Practical applications

Vertical curves are introduced wherever there is a change of gradient, be it at the crest of a hill, in a valley or at some intermediate point. Their purpose is to ease the passage from one slope to the next, minimising passenger discomfort due to radial acceleration. Also, on valley curves, vehicles tend to "bottom" on the springs, a tendency which can be reduced by decreasing the rate of change of gradient along the curve. A criterion in the design of summit curves is the sighting distance, the distance at which oncoming vehicles will first be sighted over the brow of the hill (or, in the case of railways and one-way roads, the distance at which a stationary vehicle on the other side of the brow will first be sighted). Again, the sighting distance can be increased by reducing the rate of change of gradient, i.e. by making the curve flatter.

The length of a vertical curve depends on the difference of slope of the tangents and the rate of change of gradient. Therefore the length of the curve depends on sighting distance for summit curves and dynamic considerations for valley curves. These in turn depend on the design speed and the type of road, e.g. whether it is to carry one-way or two-way traffic.

The gradient at any point of a vertical curve may be expressed as a fraction, e.g. $1/25$. It is often more convenient to express it as a percentage, however, so that $1/25$ becomes 4%. The total change of gradient, which is often required in design, can be obtained from a formula, but in the author's opinion is best calculated by common sense (see Table 9).

TABLE 9. EXAMPLES OF TOTAL CHANGE OF GRADIENT
In each, T_1 and T_2 are the tangent points.

Elevation	Change of gradient
(a)	$7 + 3 = 10\%$
(b)	$7 - 3 = 4\%$
(c)	$5 + 4 = 9\%$

SPECIMEN QUESTION 27
If the length of the curve in Table 9 (*a*) is 500 m, calculate: (*i*) the rate of change of gradient along the curve, and (*ii*) the position of the highest point on the curve.

SOLUTION

(*i*) For the parabola $y = cx^2, \dfrac{d^2y}{dx^2} = 2c = $ constant. But $\dfrac{d^2y}{dx^2}$ is the rate of gradient, which is therefore constant.

Hence, rate of change of gradient = 10% per 500 m = 1% per 50 m

(*ii*) At the highest point of the curve, the slope $\quad = 0\%$.

Change of gradient between T_1 and highest point = 7%
Since change of gradient between T_1 and $T_2 \qquad = 10\%$,
highest point occurs at a distance from $T_2 \qquad = 7/10 \times T_1T_2$
$\qquad\qquad\qquad\qquad\qquad\qquad\qquad\qquad = 7/10 \times 500$
$\qquad\qquad\qquad\qquad\qquad\qquad\qquad\qquad = 350$ m

Properties of vertical curves

Referring to Fig. 50, if:

A, C = tangent points; AB, BC = tangents; BD is perpendicular to AC, and $\therefore$ $AD = DC$; FG is parallel to BE and is a variable lying anywhere between BE and C, then the properties are:

$$BE = \tfrac{1}{2}BD$$
$$\frac{FG}{BE} = \left(\frac{CF}{CB}\right)^2$$

The properties given above can be proved true for any parabola by co-ordinate geometry.

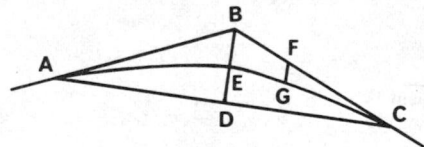

FIG. 50. *Properties of the vertical curve.*

In well-designed vertical curves the gradients should not exceed about 5%, so that uphill traffic will not be reduced to too slow a speed. For such flat curves the following approximations can be made.

(*a*) $AC = $ length, L, of the curve $AEGC$ = (chainage at C) − (chainage at A).

(*b*) Tangent length, $AB = BC = L/2$.

(*c*) (Reduced level at B) − (reduced level at E) = BE.

(*d*) (Reduced level at E) − (reduced level at D) = ED.

SPECIMEN QUESTION 28

Two slopes that intersect at chainage 8000 m at an elevation of 125 m are to be connected by a vertical curve 300 m long. If the rising slope has a gradient of 1.8% and the falling slope 1.2%, determine the elevation at each 30 m station. Find also the sight distance for drivers whose eyes are 1.1 m above road level.

SOLUTION

Referring to Fig. 51,

At A, chainage $= 8000 - \frac{1}{2} \times 300 = 7850$ m
and R.L. $= 125 - 0.018 \times 150 = 122.30$ m

At C, chainage $= 8000 + \frac{1}{2} \times 300 = 8150$ m
and R.L. $= 125 - 0.012 \times 150 = 123.20$ m

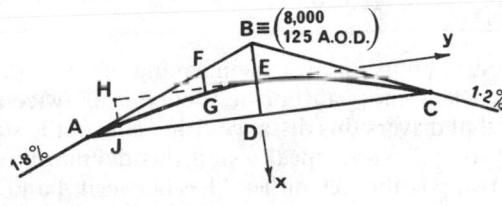

FIG. 51

The Reduced Level of D is the mean of the Reduced Levels of A and C. i.e., at D Reduced Level $= \frac{1}{2}(122.30 + 123.20) = 122.75$ m

$\therefore BD = 125.00 - 122.75 = 2.25$ m
$BE = ED = \frac{1}{2} BD = 1.125$ m, and at E, therefore,

$$R.L. = 123.875 \text{ m}$$

The reduced levels of points on the curve are calculated by deducting offsets, FG, from the reduced level of the tangent at the required chainage. Notice that the values of the offsets are symmetrical about the centre of the curve.

At chainage 8000, $FG = BE = 1.125$ m;

at chainage 7970 and 8030, $FG = (\frac{4}{5})^2 \times 1.125 = 0.720$ m;

at chainage 7940 and 8060, $FG = (\frac{3}{5})^2 \times 1.125 = 0.405$ m, and so on.

These calculations are based on the property:

$$\text{offset} \propto (\text{distance along tangent})^2$$

and are presented fully in Table 10.

TABLE 10. ELEVATIONS OF STATIONS ON CURVE
(*see Specimen Question* 28)

Chainage	R.L. on tangent	Offset (FG)	R.L. on curve
(A) 7850	122.30	0	122.300
7880	122.84	0.045	122.795
7910	123.38	0.180	123.200
7940	123.92	0.405	123.515
7970	124.46	0.720	123.740
(B) 8000	125.00	1.125	123.875
8030	124.64	0.720	123.920
8060	124.28	0.405	123.875
8090	123.92	0.180	123.740
9120	123.56	0.045	123.515
(C) 9150	123.20	0	123.200

There is some confusion over the meaning of "sight distance" but, taking $HJ = 1.1$ m, most authorities define it as twice the distance HE in order that drivers this distance either side of the summit could just see "eye to eye", so to speak—and this definition is used here.

There are two possibilities: either J lies between A and E as shown, *or J* lies below A on the straight approach. Here, assume J lies on AE.

With axes chosen as shown in Fig. 51, the equation of the parabola is $y^2 = kx$ and since $y = 150$ when $X = 1.125$,

$$k = \frac{150^2}{1.125} = 20\,000$$

When $HJ = x = 1.1$, $y^2 = 22\,000$
$$y = \sqrt{22\,000} = 148.3 \text{ m},$$

which is consistent with the assumption.

$$\therefore \text{Sight Distance} = 2 \times 148.3 = \underline{\underline{296.6 \text{ m}}}$$

NOTE

Had it been assumed that J lay below A the calculation would have been based on the properties of similar triangles thus:

$$\frac{BA}{2.25} = \frac{BJ}{1.1 + 1.125} = \frac{BJ}{2.225}$$

$$\therefore BJ = \frac{2.225}{2.250}BA, \text{ which is less than } BA.$$

This implies that J lies on BA which contradicts the assumption upon which this calculation is based and, therefore, the answer is not valid (even though the value obtained in this example by the wrong assumption is very close to the correct answer owing to the proximity, near A, of the curve AE and the straight AB).

SPECIMEN QUESTION 29

A rising gradient g_1 is followed by another rising gradient g_2 (g_2 less than g_1). These gradients are connected by a vertical curve having a constant rate of change of gradient. Show that at any point on the curve the height y above the first tangent point A is given by

$$y = g_1 x - \frac{(g_1 - g_2)x^2}{2L}$$

where x is the horizontal distance of the point from A, and L is the horizontal distance between the two tangent points.

Draw up a table of heights above A for 10-m pegs from A when $g_1 = +5\%$, $g_2 = +2\%$ and $L = 100$ m.

At what horizontal distance from A is the gradient 3%?

SOLUTION

Referring to Fig. 52, the axes are fixed by the question and pass through A, the origin.

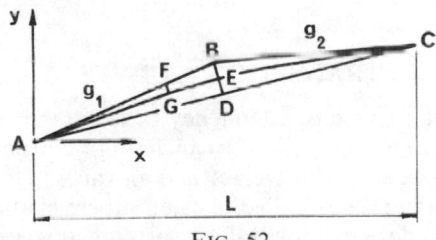

FIG. 52

Since the rate of change of gradient is constant, the curve is a parabola, and therefore the properties of the parabola given earlier in this chapter apply to this problem.

Let G be the general point on the curve, with co-ordinates (x, y), then

$$\text{Value of } y \text{ at } F = y_F = x g_1$$

and, using the same notation,

$$y_B = L/2 \cdot g_1$$
$$y_C = L/2 \cdot (g_1 + g_2)$$
$$y_D = \tfrac{1}{2} y_C = L/4 \cdot (g_1 + g_2)$$
$$\therefore\ BD = y_B - y_D = L/4 \cdot (g_1 - g_2)$$
$$\therefore\ BE = \tfrac{1}{2} \cdot BD = L/8 \cdot (g_1 - g_2)$$

$$\therefore FG = \left(\frac{x}{\frac{1}{2}L}\right)^2 BE = \frac{x^2}{2L}(g_1 - g_2)$$

$$\therefore y_G = y_F - FG = g_1 x - \frac{x^2}{2L}(g_1 - g_2)$$

This is the equation of the parabola through E, tangential to AB at A, which is the required curve. Hence (dropping the subscript $_G$) the equation,

$$y = g_1 x - \frac{x^2}{2L}(g_1 - g_2)$$

is true from A to B.

Total change of gradient $= 5\% - 2\% = 3\%$

$\therefore$ a change of 2% (i.e. from 5% to 3%) occurs in $2/3 \times$ length of the curve.

Hence the gradient is 3% at a distance from $A = 2/3 \times 100 = 66.67$ m.

The table of heights above A is left as an exercise for the reader. The values can be calculated by the method used in Specimen Question 28.

TRANSITION CURVES

A train on a flat curve has a tendency to leave the rails outwards. This tendency can be neutralised by inclining the track transversely so that a component of the weight acts inwards. The difference in height between the rails is called cant or superelevation. A circular curve of radius R designed for traffic travelling at speed V requires a certain amount of cant (which depends on V^2/R), whereas no super-elevation is required where the track is straight. It follows that for maximum safety and passenger comfort a curve which permits the gradual transition from zero superelevation to maximum super-elevation must be introduced between straight and circular curve (or, sometimes, between two circular curves of different or opposite curvatures): *see* Fig. 53.

The curve most frequently used for transitions, because it is both convenient for setting out and is a good approximation to the mathematically ideal curve (*see* Specimen Question 32) is the cubic parabola which has the equation

$$y = x^3/(6LR)$$

where $L =$ length of the transition curve and $R =$ radius of the circular curve.

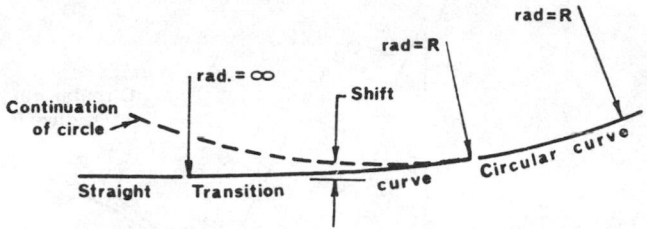

FIG. 53. *Location of a typical transition curve.*

Properties of transition curves

The results which follow are based on the approximation that δ_B is so small that the length of the transition curve may be taken as the corresponding length measured along the x axis; i.e. in Fig. 54 $TMB = TG$. Consequently, the distance along the transition curve to any point may be taken as the x co-ordinate of the point.

The order in which the properties are presented is designed to facilitate the memorising of them and is not the order in which they are logically derived. Referring to Fig. 54:

(*a*) The shift bisects the transition, and the transition bisects the shift. This means that

$$TM = MB \quad \text{(and therefore } TE = EG)$$
and
$$DM = ME$$

(*b*) The shift, $s = L^2/(24R)$. This follows from the previous result and the equation to the transition curve, $y = x^3/(6LR)$, for

$$ME = y_M = (\tfrac{1}{2}L)^3/(6LR) = L^2/(48R)$$
and
$$\text{Shift} = DE = 2ME$$

(*c*) It is easily shown that $\angle BAC = \angle GFB = \phi_B$ and, therefore, from $\triangle ABC$, $\phi_B = \sin^{-1}(BC/AB) = \sin^{-1}(\tfrac{1}{2}L/R) \simeq L/(2R)$ radians. Using the notation $BG = y_B$ and $TG = x_B$, etc.,

$$\delta_B = \tan^{-1}(y_B/x_B) = \tan^{-1}[L/(6R)] \simeq L/(6R) \text{ radians} = \phi_B/3$$

(*d*) From (*c*), $\angle TBF = 2\delta_B$.

(*e*) *Length of the transition curve*

   (*i*) This may be fixed by the geometry of the site or, in problems, it may be given.

   (*ii*) It may be governed by superelevation requirements as in Specimen Question 31.

   (*iii*) It may be determined by restricting the rate of change of radial acceleration to n m/s^3. In this case $L = V^3/(nR)$, where V is the design speed of the curve in m/s, and R is the radius in m.

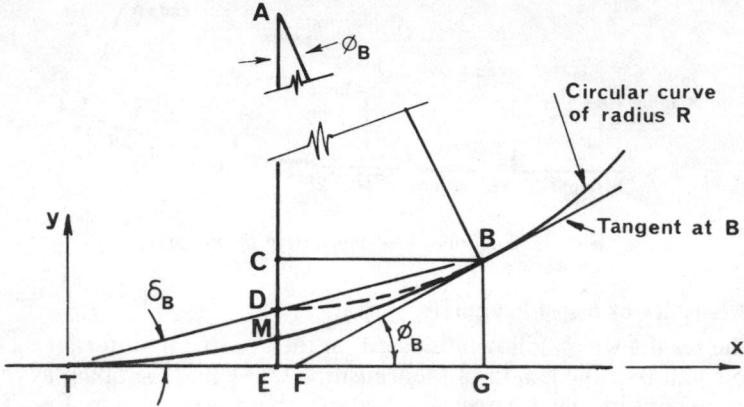

FIG. 54. *Definitions of terms associated with transition curves.*

TMB = transition curve
TG = tangent at T
T = "point of transition"
A = centre of circular curve
B = end of transition
BF = common tangent to circular and transition curves
DE = shift

Setting out a transition curve

In Fig. 54, δ_B is the setting-out angle for the point B and is usually termed the deflection angle or tangential angle. Any other point P on the curve between T and B can be set out by means of a tape and theodolite set up at T. The theodolite is used to set out the deflection angle δ_P, which is given by

$$\delta_P = \tan^{-1}(y_P/x_P) \simeq y_P/x_P \text{ radians} = x_P^2/(6LR) \text{ radians}$$

while the tape is used to establish the correct chainage of the point which, with the usual approximation, will be x_P.

The circular curve can be set out by Rankine's method of tangential angles if the theodolite is moved to B and set up there so that the telescope is pointing along the tangent at B when the horizontal scale is reading zero. In the case of a left-hand curve, as shown in Fig. 54, this is achieved by setting the horizontal scale reading to $2\delta_B$, sighting T and then, with the lower clamp fixed, turning the telescope until the horizontal scale reads $0°$ and transitting. For a right-hand curve (i.e. one of the opposite curvature) the procedure is the same except that T is sighted with the horizontal scale reading $360° - 2\delta_B$.

SPECIMEN QUESTION 30

Two tangents which intersect at an angle of 41° 40′ are to be connected by a circular curve of 1000 m radius with a transition curve at each end. The chainage of the intersection point I is 2785.7 m. The transition curves are to be of the cubic parabola type, designed for a maximum speed of 30 m/s and the rate of change of radial acceleration is not to exceed 0.3 m/s³.

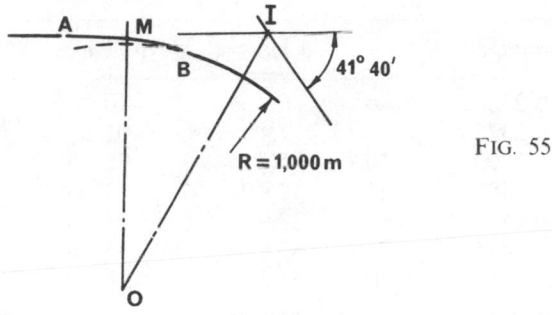

FIG. 55

Find the chainage of the beginning and end of the first transition curve and draw up a table of deflection angles to the nearest 20″ for setting out the curve in 15-m chord lengths.

SOLUTION

In Fig. 55 I is the intersection point, O is the centre of the circular arc, AB is the transition curve and M its mid-point.

Design speed, $V = 30$ m/s and rate of change of radial acceleration, $n = 0.3$ m/s³

$\therefore$ Length of transition, $L = V^3/(nR) = 27\,000/(0.3 \times 1000) = 90$ m
Shift, $s = L^2/(24R) = 90^2/24\,000 = 0.3$ m (to one decimal place)

$$\text{From Fig. 55,} \qquad \angle MOI = \tfrac{1}{2} \times 41° \, 40' = 20° \, 50'$$
$$\therefore \text{Tangent length} = AM + MI$$
$$= \tfrac{1}{2} \times 90 + (R + s) \, \tan \, 20°$$
$$= 45 + 1000.3 \tan 20° \, 50'$$
$$= 45 + 380.7$$
$$AI = 425.7 \text{ m}$$

$$
\begin{aligned}
\text{Chainage at } I &= 2785.7 \text{ m} \qquad \text{(given)} \\
\text{deduct } AI &= 425.7 \text{ m} \\
\hline
\text{Chainage at } A &= 2360.0 \text{ m} \\
\text{Add } AB &= 90.0 \text{ m} \\
\hline
\text{Chainage at } B &= 2450.0 \text{ m}
\end{aligned}
$$

For setting out, $\delta = y/x = x^2/(6LR)$ radians

i.e. $\delta = 180 \times 60/\pi \times x^2/(6 \times 90 \times 1000)$ minutes
        $= 0.006\,366\,x^2$ minutes.

From this expression the values of δ, the deflection angles, have been calculated and are given in Table 11.

TABLE 11. SETTING-OUT DATA DERIVED IN SPECIMEN QUESTION 30

Chainage (m)	x	δ mins	δ (to nearest 20″)
(A) 2360	0	0	0
2375	15	1.4	01′ 20″
2390	30	5.7	05′ 40″
2405	45	12.9	13′ 00″
2420	60	22.9	23′ 00″
2435	75	35.8	35′ 40″
(B) 2450	90	51.6	51′ 40″

NOTE

Most of the calculations have been performed on a calculator or by 4-figure logs. It should be noted however that, having calculated by 4-figure logs that $\delta = 51.56$ min (rounded off in Table 11 to 51.6 min) when $x = 90$ m, the value of δ corresponding to $x = 45$ is $\frac{1}{4} \times 51.56$ $= 12.89$ and when $x = 15$ m, $\delta = \frac{1}{9} \times 12.89$ min, these values being obtained by applying the principles of proportionality.

SPECIMEN QUESTION 31

The limiting speed around a circular curve of 600 m radius calls for a superelevation of 1/25 across the 10-m carriageway. Adopting the recommendation of a rate of 1 in 200 for the application of super-elevation along the transition curve leading from the straight to the circular curve, calculate the tangential angles for setting out the transition curve with pegs at 20-m intervals from the tangent point with the straight.

SOLUTION

Superelevation required on circular curve $= 1/25 \times 10 = 0.4$ m. As this is to be applied gradually on the transition curve at a rate of 1 in 200, its length must be $L = 0.4 \times 200 = 80$ m.

Hence, the deflection angle at 80 m from the tangent point (i.e. at the end of the transition curve) is:

$$\delta_{80} = L/(6R) \text{ radians} = 80/(6 \times 600) \text{ radians}$$
$$= 80/3600 \times 180 \times 60/\pi \text{ minutes}$$

i.e. $\delta_{80} = 76.40 \text{ minutes} = 1° \ 16' \ 24''$

The other deflection angles can be calculated by proportion, for

$$\delta_{60} = \left(\frac{60}{80}\right)^2 \delta_{80} = \frac{9}{16}\delta_{80}$$

$$\delta_{40} = \left(\frac{40}{80}\right)^2 \delta_{80} = \frac{1}{4}\delta_{80}$$

and so on, to give the following values as the answer: 04′ 46″; 19′ 06″; 42′ 58″ and 1° 16′ 24″.

The following question is the bookwork leading to the equation of the mathematically ideal transition curve referred to on p. 112. The cubic parabola equation hitherto used is obtained from this result by using the relationships between x and l, and y and l, and making certain simplifying approximations.

SPECIMEN QUESTION 32

What is the purpose of a transition curve in highway design? Derive an expression for the basic equation of a transition curve, i.e.

$$\phi = l^2/(2RL)$$

where ϕ is the angle between the straight and the tangent to the transition curve at a distance l along it from the origin; L is its total length, and R the final radius of curvature. What two considerations would each serve to determine the length L?

SOLUTION

Answers to the descriptive parts can be found in the text.
In Fig. 56:

arc $AB = L = $ length of the transition curve
P is any point of the transition curve, a distance l from A
At P, radius of curvature $= r$, and cant $= h$
$b = $ width of the carriageway
$V = $ design speed

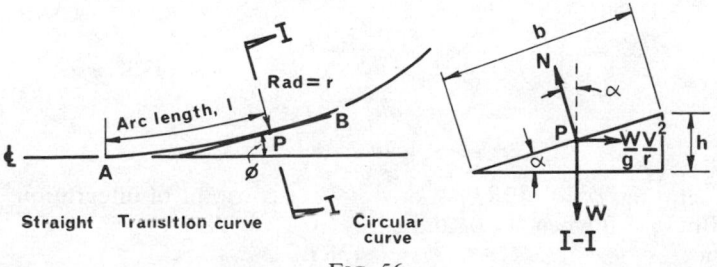

FIG. 56

Referring to section I–I, the ideal superelevation is the value of h for which the three forces shown acting at P are in equilibrium.

i.e.
$$\tan \alpha = \frac{WV^2}{gr} \bigg/ W = \frac{V^2}{gr}$$

But, also, $\qquad \sin \alpha = h/b$

For small values of $\alpha \simeq \tan \alpha$

Hence, providing the superelevation is small,

$$h/b = V^2/gr;$$

hence $\qquad hr = \text{constant}$ . . . . . (1)

Maximum rates of application of superelevation have been recommended and as the cost per unit length will be less for straights than for curves, it is desirable to keep the curve as short as possible, namely, by applying the superelevation at the maximum rate (which is constant).

It follows that

$$h \propto l \qquad . \qquad . \qquad . \qquad . \qquad . \qquad (2)$$

From (1) and (2), $\qquad rl = \text{constant}.$

At B, $r = R$ and $l = L$ and, since the constant has the same value at all points on the transition curve, including B,

$$rl = RL. \qquad . \qquad . \qquad . \qquad . \qquad . \qquad (3)$$

The variables l and ϕ are, in fact, the intrinsic co-ordinates of the transition curve. Reference to a suitable book on mathematics will show that the radius of curvature is defined in terms of the intrinsic co-ordinates by

$$r = \frac{dl}{d\phi}$$

Substituting for r in (3),

$$l\frac{dl}{d\phi} = RL$$

i.e.
$$\times \frac{d\phi}{dl} = \frac{l}{RL}$$

Integrating, $\phi = l^2/(2RL) + k,$ $\qquad k = \text{constant of integration}.$

But $\phi = 0$ when $l = 0$, and $\therefore k = 0$,

hence $\qquad \underline{\underline{\phi = l^2/(2RL)}}$

EXAMINATION QUESTIONS

1. An uphill gradient of 1 in 100 meets a downhill gradient of 0.44 in 100 at a point where the chainage is 6100 m and the reduced level is 126 m. If the rate of change of gradient is to be 0.18% per 30 m, prepare a table for setting out a connecting vertical curve at intervals of 30 m.

2. A falling gradient of 1 in 60 is followed by a rising gradient of 1 in 120. A vertical parabolic curve to join the two gradients must have a level of 196.25 m above datum at the mid-point to give the necessary clearances from existing drainage and overhead structures. Find the length of the curve and the levels at 30-m intervals if the two gradients meet at a level of 195.50 m above datum, at a chainage of 3400 m. Find the chainage and level at the lowest point on the curve.

3. Show from first principles, that when a vertical parabolic curve of the summit variety is to be inserted on a new road to connect an upwards gradient of a per cent to a downward gradient of b per cent,

(i)
$$L = \frac{S^2(a + b)}{800h}$$

when the length of the curve $(\overline{L})$ is greater than the sight distance and h is the height of the driver's eye above the road or

(ii)
$$L = 2S \frac{800h}{(a + b)}$$

when the length of the curve (L) is less than the sight distance h again being the height of the driver's eye above the road.

State any assumptions which are made in the derivation of the formulae.

4. In a road construction scheme, the point T_1 (chainage 330.00 m and reduced level 77.300 m A.O.D.) lies on an up gradient of 1 in 40. T_1 is the start of a parabolic vertical curve T_1T_2, 520 m long, which connects the up gradient to a down gradient, the road reaching a summit between T_1 and T_2. The reduced level of the road surface at a chainage of 530.00 is 79.800 m. Determine:

(a) the chainage and reduced level of the road surface at T_2;
(b) the gradient at T_2;
(c) the surface level at a chainage of 700 m;
(d) the minimum sighting distance over the summit, assuming eye heights of 1.1 m above road level.

5. Briefly outline and discuss the criteria for determining the length of a valley vertical road curve.

A valley vertical curve is to be constructed connecting a down gradient of 1 in 50 to an up gradient of 1 in 25. A typical vehicle using the curve has headlights 0.76 m above the road surface; the beam being inclined upwards 1° to the horizontal. At night the headlights are to illuminate the road at least 100 m ahead of the vehicle. Determine:

(a) the curve length;

(b) the chainage and level of the lowest point on the road if the chainage and level of the first tangent point on the 1 in 50 slope are 2983.68 m and 20.628 m respectively;

(c) the chainage and level of the tangent point on the 1 in 25 slope.

6. Why are highway transition curves introduced? Explain the types of curve that may be used, and give their limits of application.

A curve is to be introduced between two straights of a motorway that deflect 55°. Given that the maximum allowable rate of application of superelevation is 1 in 200; the maximum allowable superelevation 1 in $14\frac{1}{2}$, the design road speed 112.7 km/h, and carriageway width 10 m. Determine:

(a) the length and radius of a circular curve in the middle of the two transitions connecting these straights;

(b) the data necessary to set out the first three points on the first transition to an adequate degree of accuracy.

7. A road curve of 2 km radius joins two intersecting straights tangentially at chainages 3689.52 m and 4392.76 m.

A new road curve with transitions is to be constructed to link the straights, the specification being as follows:

Design speed 100 km/h
Maximum superelevation 1 in $14\frac{1}{2}$
Allowable rate of gain of centripetal acceleration 0.3 m/s^3

For the new road determine:

(a) the circular curve radius;

(b) length of transitions;

(c) the data to set out one of the transitions at 10 m chainage points;

(d) the chainage of the final tangent point on the straight.

8. A railway curve of 1000 m radius is to be set out with transition curves at each end to connect two straights which have a total deflection angle of 35° 30′. The chainage of the point of intersection is 1465.2 m. The transition curve is to be of the cubic parabola form, designed for a maximum speed of 30 m/s given that the rate of change of radial acceleration must not exceed 0.3 m/s^3.

Determine the chainage at the beginning and end of the first transition curve and tabulate the deflection angles required for setting out the curve at chord lengths of 30 m. What would be the amount of superelevation on the circular curve?

9. It is required to set out, using 15-m chords, a circular curve of radius 500 m linked to a given straight line by a cubic transition curve. The transition AB is to have a rate of change of radial acceleration of 0.5 m/s^3 when traversed at 100 km/h.

(*i*) If the starting-point of the transition is at chainage 1400 m, find the chainage at B, the start of the circular curve.

(*ii*) Where and how would the theodolite be set up for setting out the circular curve?

(*iii*) Tabulate data for setting out the first three pegs (excluding B) on the circular curve if the chainage is to be "through". Give deflection angles to the nearest 20 seconds.

10. BA and AC are two straight lines on a traverse for a proposed railway, AC deflecting 30° to the right from BA at A. It is proposed to join these lines by a circular curve of 800 m radius with a cubic parabola transition at each end. The maximum speed for which the curve is to be designed is 20 m/s, and the length of each transition is to be such that the cant is applied at the rate of 25 mm per second. Determine:

(*a*) the distance of the first tangent point T_1 (on BA) from A;

(*b*) the total length of the curve (transition–circular arc–transition) to T_2;

(*c*) the right-angle offsets from T_1A to the transition, the pegs to be placed at 20-m intervals from T_1 (to the nearest 5 mm).

Gauge of track between centre of rails = 1.5 m.

Acceleration due to gravity = 9.81 m/s^2.

[*Hint.* Offsets y corresponding to chainages x can be calculated from the equation to the cubic parabola, $y = x^3/(6LR)$.]

CHAPTER 8

APPLICATIONS OF PLANE TRIGONOMETRY

Sometimes examination questions in surveying are found which reduce to fairly simple trigonometry problems. A selection of such examples is given in this chapter; by working through them it will be seen that, apart from knowing the definitions of a few surveying terms, their solutions depend on knowing only the following elementary relationships:

(a)
$$\frac{\sin A}{a} = \frac{\sin B}{b} = \frac{\sin C}{c}$$

where A, B and C are the angles of the triangle ABC and a, b and c are, respectively, the sides opposite these angles.

Using the same notation:

(b)
$$a^2 = b^2 + c^2 - 2bc \cos A$$

and the two similar formulae for b^2 in terms of $\cos B$, and c^2 in terms of $\cos C$.

(c) Area of triangle $ABC = \sqrt{s(s-a)(s-b)(s-c)}$

where $s = \frac{1}{2}(a + b + c)$.

(d) Area of triangle $ABC = \frac{1}{2}ab \sin C = \frac{1}{2}bc \sin A = \frac{1}{2}ca \sin B$

(e) The approximation that, for small angles, $\theta \simeq \sin \theta \simeq \tan \theta$.

Reference to the appropriate mathematical tables will show that, for:

7-figure accuracy, θ radians $= \sin \theta$ if $\theta \leqslant 18$ minutes

θ „ $= \tan \theta$ if $\theta \leqslant 21$ „

6-figure accuracy, θ „ $= \sin \theta$ if $\theta \leqslant 45$ „

θ „ $= \tan \theta$ if $\theta \leqslant 40$ „

4-figure accuracy, θ „ $= \sin \theta$ if $\theta \leqslant 4$ degrees

θ „ $= \tan \theta$ if $\theta \leqslant 3° 23'$

An extension of this approximation is that, for example,

$$\sin 0° 04' = 4 \sin 1' = 240 \sin 1''$$
or
$$\log \sin 0° 04' = \log 240 + \log \sin 1''$$

Some surveyors remember the relationships:

(a) 206 265 seconds = 1 radian
(b)   log sin $1'' = \log 1/206\,265 = \bar{6}.685\,575$

SPECIMEN QUESTION 33
The length of a base-line is to be determined by measuring the angle subtended at one end of the line by a 2-metre invar subtense bar set up at the other end. This angle is found to be $0°\ 16'\ 40'' \pm 1$ second of arc.

Determine the length of the base-line and, assuming that the subtense bar, being invar, is exactly 2 metres between targets, express the accuracy of the length so determined as a fraction ($\pm 1/N$).

What is the accuracy of the length of the base, as the sum of two halves, determined by measuring the angles subtended at each end by the subtense bar set up at the mid-point?
             (1 radian = 206 265 seconds of arc)

SOLUTION
In Fig. 57 (a), AB represents the subtense bar = 2 m long.

     CD represents the base-line of length L metres, say.

A theodolite is stationed at C to measure the angle of $0°\ 16'\ 40''$ subtended there by the targets at A and B each end of the subtense bar.

Since $0°\ 16'\ 40'' = 1000'' = 1000/206\,265$ radians, which is very small, AB can be considered an arc of a circle of radius L, so

$$2 = L \times 1000/206\,265$$
$$\therefore L = 412.53 \text{ m}$$

Since the angle is measured to within ± 1 second of arc, the accuracy is $\pm 1/1000$.

Hence, $\underline{\underline{L = 412.53, \pm 1/1000}}$

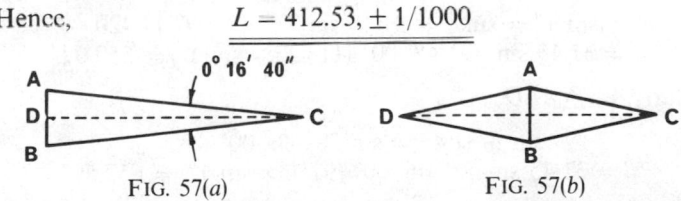

FIG. 57(a)   FIG. 57(b)

In Fig. 57(b), AB = subtense bar
          CD = base-line, C and D being theodolite stations

If AB is exactly midway between C and D, the angles subtended at C and D will be $2 \times 0°\ 16'\ 40''$. Using the same theodolite, both angles can again be measured to within ± 1 second of arc. Hence, accuracy of each half will be = 1/2000.

Now, one result of the theory of errors, *see* p. 203, is that if $A = B + C + D$, B, C and D being measured quantities, then

$$e_A = \sqrt{e_B{}^2 + e_C{}^2 + e_D{}^2}$$

where e_A, e_B, e_C and e_D are the probable errors in A, B, C and D respectively. As this result is true for any number of terms on the right-hand side, the error in the base-line measurement

$$= \sqrt{\left(\frac{1}{2000} \cdot \frac{L}{2}\right)^2 + \left(\frac{1}{2000} \cdot \frac{L}{2}\right)^2}$$

$$= \frac{L}{2828}$$

$$\therefore \text{ Accuracy of base-line computation} = \pm \frac{1}{2828}$$

SPECIMEN QUESTION 34

The details given below refer to observations made at a satellite station O, in order to determine the angle at an inaccessible station A in a triangle ABC. Compute the angle BAC.

Length $OA = 31.45$ m	Bearing of side $OA = 00°\ 00'\ 00''$	
„ $\quad AB = 9750$ m	„ $\quad$ „ $\quad OB = 78°\ 46'\ 00''$	
„ $\quad AC = 11\,420$ m	„ $\quad$ „ $\quad OC = 100°\ 12'\ 00''$	
Log sin $1'' = \overline{6}.685\,575$		

SOLUTION

In Fig. 58, let $\qquad \angle ABO = \beta$ seconds

$$\angle ACO = \gamma \qquad \text{„}$$

In $\triangle ACO$, $\sin \gamma / 31.45 = \sin 100°\ 12'\ 00'' / 11\,420$

$$= \sin 79°\ 48'\ 00'' / 11\,420$$

$$\therefore \gamma \sin 1'' \simeq \sin \gamma = 31.45 \sin 79°\ 48'\ 00'' / 11\,420$$

i.e. $\quad \gamma = 31.45 \sin 79°\ 48'\ 00'' / (11\,420 \times \sin 1'') = 559.07''$

Similarly, in $\triangle ABO$,

$$\sin \beta / 31.45 = \sin 78°\ 46'\ 00'' / 9750$$

and $\quad \beta = 31.45 \sin 78°\ 46'\ 00'' / (9750 \times \sin 1'') = 652.59''$

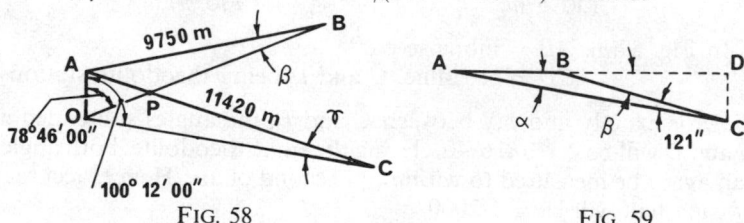

FIG. 58   FIG. 59

In Fig. 58,

$$\angle BPC = \angle BOC + \gamma = 21° \ 26' \ 00'' + \gamma$$

also $\qquad \angle BPC = \angle BAC + \beta$

$$\therefore \ \angle BAC = 21° \ 26' \ 00'' + \gamma - \beta$$
$$= 21° \ 26' \ 00'' + 09' \ 19''.07 - 10' \ 52''.59$$
$$= 21° \ 24' \ 26''.48$$

To the nearest whole second,

$$\underline{\underline{\angle BAC = 21° \ 24' \ 26''}}$$

NOTES

(*i*) 6-figure logs are used in (and are necessary for) these calculations.

(*ii*) The use of a satellite station, O, would be necessary in practice if, for example, the signal A sighted from B and C were part of a church steeple, in which case it would be impossible to set up a theodolite at A.

SPECIMEN QUESTION 35

Two plumb wires A and B in a shaft were 7.400 m apart. A theodolite was set up at C slightly off the line AB at a distance of 12.527 m from the wire B, and nearer B than A. The angle ACB was found to be 121 seconds.

Calculate the perpendicular distance from C to AB produced

$$(\log \sin 1'' = \bar{6}.6856).$$

SOLUTION

Let $\angle BAC = \alpha$ and $\angle DBC = \beta$ (*see* Fig. 59). These are the angles to be determined.

In $\triangle ABC$, $\qquad AB = 7.400$ m and $BC = 12.527$ m

$$\sin \alpha = 12.527 \sin 121''/7.400, \text{ by the sine formula.}$$
$$\alpha \sin 1'' = 12.527/7.400 \times 121 \sin 1''$$
$$\therefore \ \alpha = 12.527 \times 121/7.400 = 204.4'', \text{ using 4-figure logs}$$

Hence, $\qquad \beta = \alpha + 121'' = 325.4$

$$\therefore \ CD = 12.527 \sin \beta = 12.527 \times 325.4 \times \sin 1''$$

Using 4-figure logs,

$$CD = 0.01976 \text{ m}$$
$$\underline{\underline{= 19.8 \text{ mm}}}$$

NOTE

This example illustrates a technique used in tunnel surveying which, in all its aspects, is described clearly and concisely in *Higher Surveying*, by A. L. Higgins (Macmillan & Co.).

EXAMINATION QUESTIONS

1. Distinguish between mistakes, cumulative errors and compensating errors in chaining, giving one example of each.

A quadrilateral *ABCD* with a diagonal *AC* is part of a chain survey. Two 30-m chains, *X* and *Y*, are used, the measured lengths of the lines being as follows:

Line	Measured length (metres)	Chain used
AB	96.3	*X*
BC	118.7	*X*
CD	154.8	*Y*
DA	93.6	*Y*
AC	173.7	*Y*

After the lines had been measured it was found that the chains were of incorrect length when tested against a standard tape, their errors being: chain *X*, 100 mm too long, and chain *Y*, 50 mm too short. Determine the true area in hectares of *ABCD*. (Work to the nearest 0.1 m for lengths.)

[Note. Mistakes occur when the surveyor is at fault, e.g. by incorrectly booking readings; cumulative errors are errors which will always be of the same sign, e.g. the incorrect chain lengths in this question; compensating errors are those which tend to cancel, such as careless holding of the chain.]

2. *X* and *Y* are inaccessible points on either side of a deep gorge, and it is required to determine the horizontal distance between them. A straight base line *AB* is laid out in the valley and after all the necessary corrections have been made, the length *AB* is estimated to be 453.6 m (to the nearest 0.1 m). A theodolite is set up at each end of the base-line in turn and angles *BAX*, *ABX*, *BAY* and *ABY* measured, their values being as follows (to the nearest minute):

$$\text{Angle } BAX = 68° \ 42'$$
$$ABX = 52° \ 24'$$
$$BAY = 37° \ 50'$$
$$ABY = 45° \ 10'$$

Calculate the horizontal length of XY to the nearest 0.1 m.

What precautions would be taken in measuring these angles?

3. Describe how you would transfer a surface bearing down a shaft and set out a line underground in the same direction.

Two plumb lines A and B in a shaft are 8.24 m apart and it is required to extend the bearing AB along a tunnel. A theodolite can only be set up at C 19.75 m from B and a few millimetres off the line AB produced. If the angle BCA is 09′ 54″ what is the offset distance of C from AB produced?

4. (*a*) In a triangle FGH the side FG has been measured as 3020.60 m and the angle FHG as $65°\ 17′\ 40″$.

From F and G a mark J, 5.224 m from H, is observed, the measured angles being

$$JFG = 48°\ 10′\ 20″ \quad \text{(Note } JFG > HFG$$
$$JGF = 66°\ 29′\ 22″ \quad \text{and } JGF < HGF)$$

The angle JHG is also measured as $110°\ 17′\ 40″$.

Determine to the nearest second all the angles of the triangle FGH.

(*b*) What practical circumstances might give rise to this problem?

CHAPTER 9

EARTHWORKS

It is difficult to think of any construction work in which the removal of earth, or the dumping of earth, is not involved. Sometimes the quantities of earth are vast, the earthworks then accounting for a large proportion of the cost. In such cases it is obviously desirable that volumes should be estimated as accurately as possible. When, therefore, the shape of the excavation (or fill) can be subdivided into known geometrical shapes (pyramids are common in problems), its volume should be calculated by the relevant formula.

For a pyramid, $\qquad$ Volume $= \frac{1}{3} \times$ Base $\times$ Height

SPECIMEN QUESTION 36

Fig. 60 (a) shows the commencement of a cutting. The side slopes are $1\frac{1}{2}$ horizontal to 1 vertical, while the cross slope is everywhere 4 horizontal to 1 vertical. Calculate the volume of excavation in cubic metres.

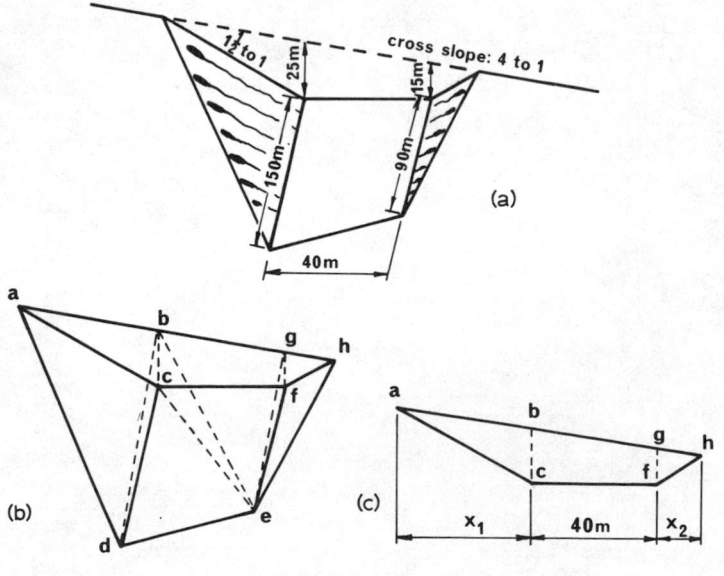

FIG. 60

SOLUTION

Fig. 60 (b) shows the cutting subdivided into four pyramids. In order to calculate the volumes of pyramids abcd and efgh, it is first necessary to calculate x_1 and x_2 (see Fig. 60 (c)) thus.

In $\triangle abc$,   Height of a above $b = x_1/4$ (from cross slope)

$\therefore$ Height of a above $c = x_1/4 + 25$

Also,   Height of a above $c = 2/3 . x_1$ (from slope of ac)

Hence, $2/3 . x_1 = x_1/4 + 25$,   from which

$$x_1 = 60 \text{ m}$$

Similarly, from $\triangle fgh$, $2/3 . x_2 = 15 - x_2/4$

from which, $x_2 = 180/11 \text{ m}$

The volumes of the four pyramids are as follows:

Pyramid		
Pyramid $abcd = 1/3 \times 25 \times 60/2 \times 150$	$= 37\,500 \text{ m}^3$
„ $bcde = 1/3 \times 150 \times 40/2 \times 25$	$= 25\,000$ „
„ $bcefg = 1/3 \times 40(15 + 25)/2 \times 90$	$= 24\,000$ „
„ $efgh = 1/3 \times 15/2 \times 180/11 \times 90$	$= 3\,680$ „

Total Volume   90 180 m³

AREAS

In many problems it is necessary to compute an area before the volume can be calculated and, for this reason, the main methods of determining areas are summarised below.

(a) The areas of the most common regular shapes are summarised in Fig. 61.

(b) *Trapezoidal rule*

Area $= \frac{1}{2}d(y_1 + 2y_2 + 2y_3 + \ldots\ldots 2y_{n+1} + y_n)$ where $y_1, y_2 \ldots\ldots$ y_n are ordinates, equally spaced a distance d apart, *see* Fig. 62 (a).

(c) *Simpson's rule*

Area $= \frac{1}{3}d(y_1 + 4y_2 + 2y_3 + 4y_4 + 2y_5 + \ldots + 2y_{2n-1} + 4y_{2n}$ $+ y_{2n+1})$

where $y_1, y_2 \ldots$ etc., are ordinates, equally spaced a distance d apart, *see* Fig. 62 (b). Please note that there has to be an *odd* number of ordinates. In other words there has to be $(2n + 1)$ ordinates where n can be 1 or any other whole number. This formula is considered to be more accurate than the less cumbersome trapezoidal rule.

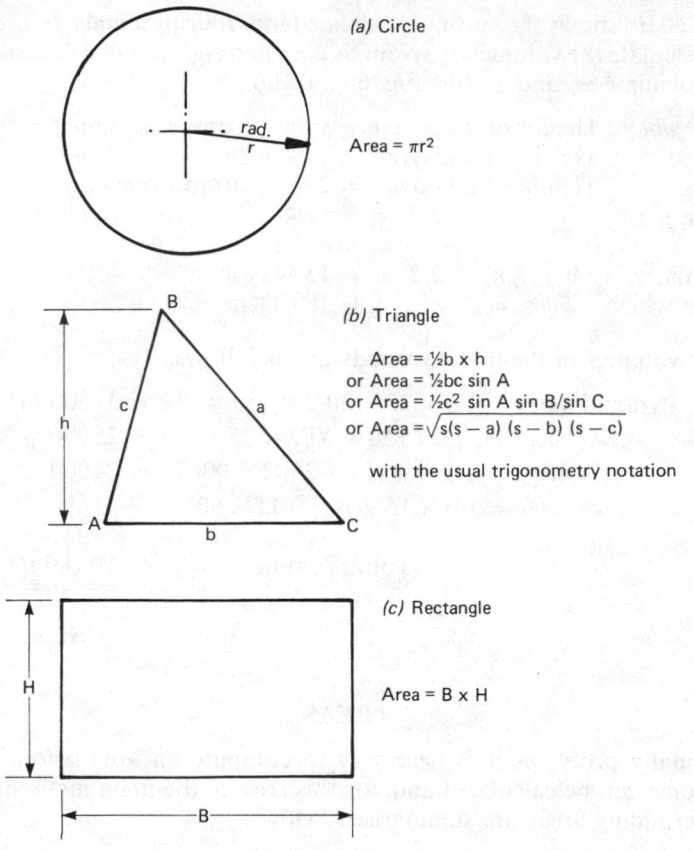

(a) Circle

Area = πr^2

(b) Triangle

Area = ½b x h
or Area = ½bc sin A
or Area = ½c² sin A sin B/sin C
or Area = $\sqrt{s(s-a)(s-b)(s-c)}$

with the usual trigonometry notation

(c) Rectangle

Area = B x H

FIG. 61. *Common formulae for area.*

(d) Mechanical methods

If a shape is plotted to scale on graph paper, its area can be measured either by planimeter or by square-counting.

SPECIMEN QUESTION 37

A plot of land is bounded by three straight fences BC, CD, and DA and an irregular hedge which lies entirely outside the quadrilateral $ABCD$. The following observations are taken:

$AB = 212.4$ m, $BC = 201.6$ m, $CD = 231.0$ m, $DA = 198.4$ m, $AC = 237.8$ m.

Offsets are taken from AB to the irregular hedge as follows:

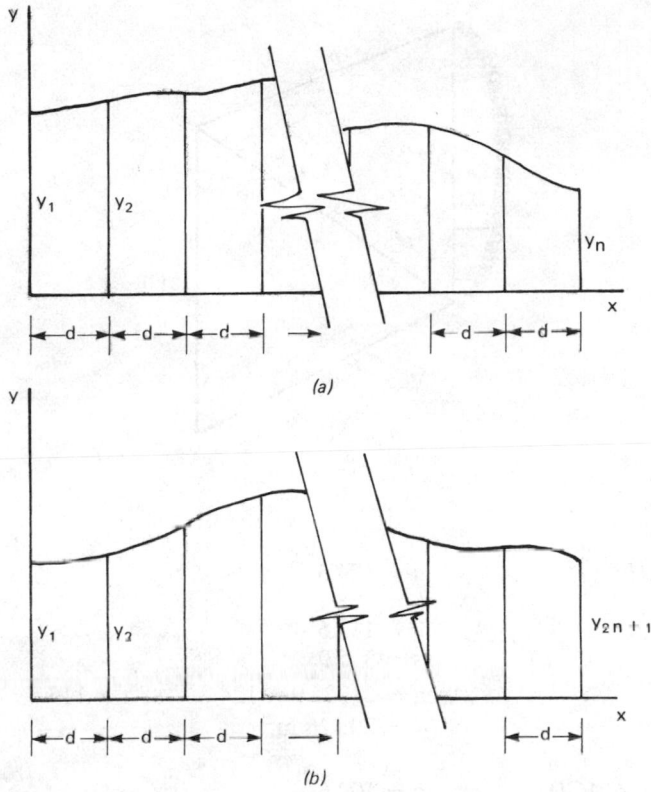

FIG. 62. (a) Diagram for trapezoidal rule. (b) Diagram for Simpson's rule.

Point (m)	$A(0)$	40	80	120	160	200	$B(212.4)$
Offset (m)	0	3.2	8.4	11.7	13.3	5.1	0

Calculate the total area of the plot, using for the irregular area (i) Simpson's rule and (ii) the trapezoidal rule.

Comment on the difference between the overall area determined by the two methods.

SOLUTION
See Fig. 63.
First calculate the semi-perimeter, $s = \frac{1}{2}(a + b + c)$
For $\triangle ABC$,

$$s = \tfrac{1}{2}(201.6 + 237.8 + 212.4) = 325.9$$

For $\triangle ACD$,

$$s = \tfrac{1}{2}(231.0 + 198.4 + 237.8) = 333.6$$

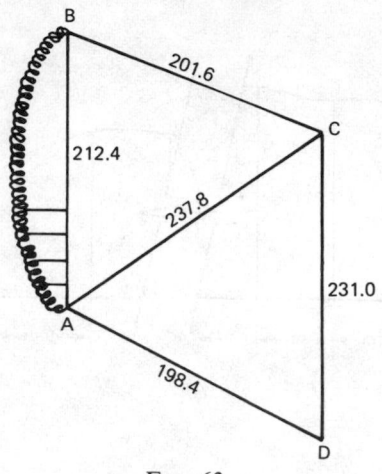

FIG. 63

Hence:

in $\triangle ABC$

$$s - a = 124.3$$
$$s - b = 88.1$$
$$s - c = 113.5$$
$$s = 325.9$$

$\therefore$ Area $= \sqrt{325.9 \times 124.3 \times 88.1 \times 113.5}$

             $= 20\,126 \text{ m}^2$

and in $\triangle ACD$

$$s - a = 102.6$$
$$s - c = 135.2$$
$$s - d = 95.8$$
$$s = 333.6$$

$\therefore$ Area $= \sqrt{333.6 \times 102.6 \times 135.2 \times 95.8}$

             $= 21\,055 \text{ m}^2$

$\therefore$ combined area, $\triangle ABC + \triangle ACD = 41\,181 \text{ m}^2$

Additional Area

By Simpson's rule $= \dfrac{40}{3} \left\{ 4(3.2 + 11.7 + 5.1) + 2(8.4 + 13.3) \right\}$

                 $= 1645 \text{ m}^2$

By trapezoidal rule $= \dfrac{40}{2} \left\{ 0 + 0 + 2(3.2 + 11.7 + 5.1 + 8.4 + 13.3) \right\}$

                  $= 1668 \text{ m}^2$

Adding the combined area of 41 181 m^2 to each of these in turn gives

using Simpson,   total area = 42 826 m^3, say <u>42 800 m^2</u>

using trapezoidal,   „   „   = 42 849 m^3, say <u>42 800 m^2</u>

Hence, to three significant figures, the answer is the same using both methods. In fact, the discrepancy between the two answers is approximately $\dfrac{2}{4280} \simeq 0.047\%$. Since 3-figure accuracy is reasonable for this type of calculation, it does not matter which of the two rules is used in this calculation.

This is not always the case; it applies here because the irregular area is only a small part of the whole area and the discrepancy between the Simpson area and the trapezoidal area is, so to speak, diluted by the much larger area to which they are added.

In point of fact the discrepancy between the Simpson area and the trapezoidal area as calculated is about 1.4%.

PRISMOIDAL AND END AREAS FORMULAE

Prismoidal formula

For the purpose of this paragraph a truncated pyramid is defined as having its end parallel, a prism is a solid object of constant cross-section with parallel ends, and a prismoid is a solid object with plane sides and parallel ends. Although one end of a prismoid may, for example, be five-sided and the other end six-sided, it is always possible to break a prismoid down into a number of pyramids, truncated pyramids and prisms—a good exercise for the imagination! It can be proved mathematically that, for any of the above-mentioned shapes:

$$\text{Volume} = h/3 \,.\, (a_1 + 4a_2 + a_3), \quad \text{exactly}$$

where a_1 and a_3 are the areas of the end planes and a_2 is the cross-sectional area of the plane midway between them. $2h$ is the height of the solid, so h is the normal distance between adjacent planes (a_1 and a_2, or a_2 and a_3).

Evidently, the volume of a prismoid $= h/3 \,.\, (a_1 + 4a_2 + a_3)$ exactly. The volume of a second prismoid of the same height $2h$ and with areas a_3, a_4 and a_5 corresponding to a_1, a_2 and a_3 will be

$$\text{Volume} = h/3 \,.\, (a_3 + 4a_3 + a_5)$$

Similarly for a third prismoid, volume $= h/3 \,.\, (a_5 + 4a_6 + a_7)$, and so on.

Adding these results gives the prismoidal formula in its general form:

$$\text{Volume} = h/3 \, . \, (a_1 + 4a_2 + a_3) + h/3 \, . \, (a_3 + 4a_4 + a_5)$$
$$+ \, h/3 \, . \, (a_5 + 4a_6 + a_7)$$

i.e. $\text{Volume} = h/3 \, . \, (a_1 + 4a_2 + 2a_3 + 4a_4 + 2a_5 + 4a_6 + a_7)$

which gives the volume of a solid bounded by plane sides in terms of the areas of (in this case) seven cross-sections and the distance, h, between adjacent sections. Obviously, the formula (which is of exactly the same form as Simpson's rule for areas) can be extended two terms at a time, so that there will always be an odd number of terms in the brackets.

The restriction that the cross-sections must be parallel is often relaxed slightly, the formula being sufficiently accurate if the centre-line of the solid forms a gentle curve. Where necessary (*see* Specimen Question 42) a correction for a curved centre-line can be applied.

End areas method

A simpler formula for the same volume is given by the end areas formula, namely,

$$\text{Volume} = \tfrac{1}{2}h(a_1 + 2a_2 + 2a_3 + 2a_4 + 2a_6 + a_7)$$

which has the advantage that it can be applied to any number of sections, odd or even. This formula is exact, however, only if all the sections are of the same constant width. By taking sufficient sections, however, any degree of accuracy can be obtained, as can be demonstrated by calculating the volume of a right circular cone of height H and base area A, viz.:

2 sections: volume $= \tfrac{1}{2}H(A + 0) = \tfrac{1}{2}AH = 0.5AH$

3 sections: volume $= \tfrac{1}{2}H/2 \, . \, (A + 2 \, . \, A/4 + 0) = 3/8 \, . \, AH$
$$= 0.375AH$$

5 sections: volume $= \tfrac{1}{4}H/2(A + 2 \times 9/16 \, . \, A + 2 \times \tfrac{1}{4}A$
$$+ \, 2 \times A/16 + 0)$$
$$= 0.344AH,$$

Similarly, for 9 sections, volume $= 0.336AH$, etc.

The exact value for the volume of a cone, $AH/3$, is well known. It will be found that the prismoidal formula gives this value no matter how many sections are taken—provided the number is odd.

SPECIMEN QUESTION 38
In order to find the excavation required for a railway cutting, cross-sections are taken at every 30 m. As the ground surface is very irregular, the cross-sections are plotted and their areas obtained by planimeter, the results being as follows:

Chainage of section (m): 1830, 1860, 1890, 1920, 1950,
Area in m²: 24,   196,   248,   101,   197,
Chainage (m): 1980, 2010, 2040, 2070, 2100, 2130.
Area in m²: 296, 362, 311,   89,   143,   49.

(a) Calculate the volume of excavation in m³ using both the end areas formula and the prismoidal formula.

(b) State which of the above answers you consider to be the more accurate, giving your reasons.

SOLUTION

(a) The prismoidal formula may be memorised as follows:

Volume = $h/3$ × (first + last ordinates, + 4 × even ordinates,
$$+ 2 × \text{odd ordinates}).$$

This form provides the key to a convenient way of writing it down, to reduce mistake-liability, thus:

$$\text{Volume} = \frac{30}{3}\begin{bmatrix} & 196 & 248 \\ 24 & + 101 & + 197 \\ 49 & 296 & 362 \\ — & 311 & 89 \\ 73 & 143 & — \end{bmatrix}$$

$$2 × 896$$

Adding the terms in the brackets: 4 × 1047   73
4 × 1047 = 4188
2 × 896 = 1792
──────
6053

By the prismoidal formula, Volume = 30/3 × 6053 m³ = 60 530 m³

By the trapezoidal formula,
Volume = 30/2 . [24 + 49 + 2(1047 + 896)] = 30/2 × 3959 m³
= 59 385 m³

(b) The prismoidal formula yields the more accurate answer for the reasons given on p. 134.

NOTE

In calculating the volume by the trapezoidal formula the chance of arithmetical mistakes would have been reduced if the addition had again been performed "vertically" instead of "horizontally".

Another common application of the *End Areas* method is the calculation of the capacity of a proposed natural reservoir from a contoured plan of the valley. Specimen Question 39 is an example.

Notice that to determine volumes corresponding to intermediate depths (i.e. between contours) it is necessary to plot volume against depth, *see* Fig. 64.

SPECIMEN QUESTION 39

The site of a proposed new reservoir is illustrated in Fig. 64 (*a*). The area enclosed by each of the various contour lines and the vertical dam is as follows:

Contour (m)	Area enclosed (m²)	Contour (m)	Area enclosed (m²)
460	31 000	475	147 000
465	87 000	480	205 000
470	106 000	485	240 000

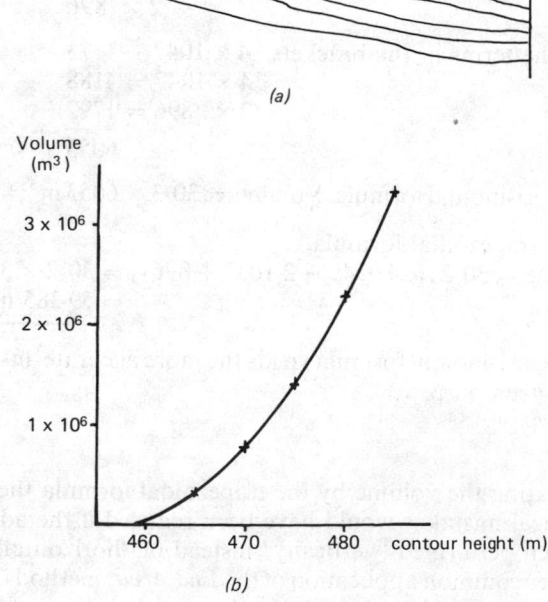

FIG. 64. (a) See *Specimen Question 39* (b) *Plot of volume* v_s *contour height.*

Calculate:

(*i*) the capacity of the reservoir if the top water level is 482 m above datum; and

(*ii*) the height above datum of the water surface when the reservoir is half full.

SOLUTION

Contour	Area (m²)	Volume increment	Volume (m³)
460	31 000		
465	87 000	295 000	295 000
470	106 000	482 000	777 000
475	147 000	633 000	1 410 000
480	205 000	880 000	2 290 000
485	240 000	1 112 000	3 402 000

In the above table the volume increment is calculated by applying the end-area method to successive areas.

The lowest point on the bed of the reservoir is below 460 *AD*. The volume of this depression has been ignored; to check if this is reasonable, the depth is estimated from the graph, Fig. 64 (*b*) to be about 1 m and, therefore,

$$\text{ignored volume} \simeq \tfrac{1}{2} \times 1(0 + 31\,000) \simeq 15\,000 \text{ m}^3.$$

From Fig. 64 (*b*).

Volume to 482 *AD* = $\underline{\underline{2.7 \times 10^6 \text{ m}^3}}$

Depth when half full, i.e. when $V = 1.35 \times 10^6$, is: $\underline{\underline{474.5 \ AD}}$

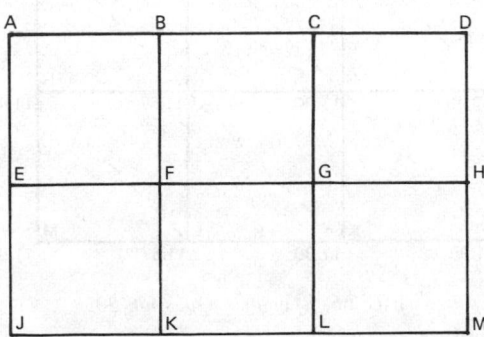

FIG. 65. *Method of "spot heights"*.

Method of spot heights

If spot levels are taken at A, B, ... L, M, which constitute a level grid, *see* Fig. 65 then they can be used to determine the volume of an excavation, with vertical sides, down to a known level.

The volume of the element $ABFE$, says, is:

its plan area × average depth at A, B, F and E

Similarly the volume of $BCGF$ = plan area × average depth at B, C, G and F. The depth at B appears in both calculations, as will the depth at C. Since four squares meet at F, the depth at F will appear in four calculations ... and so on. Provided that each of the elements $ABFE$, $BCGF$, etc., is the same in plan, it follows that the excavated volume will be given by

plan area $ADMJ$ × mean of (1 × depth at A + 2 × depth at B
$+ 4 \times$ depth at $F + ...$)

and this calculation can best be set out in a table as demonstrated in Specimen Question 40.

SPECIMEN QUESTION 40

The diagram Fig. 66 shows the existing ground levels on a 15 m square grid forming part of a site which is to be excavated to a uniform formation level of 10.00 m above datum.

Calculate the volume of earth to be excavated assuming vertical sides.

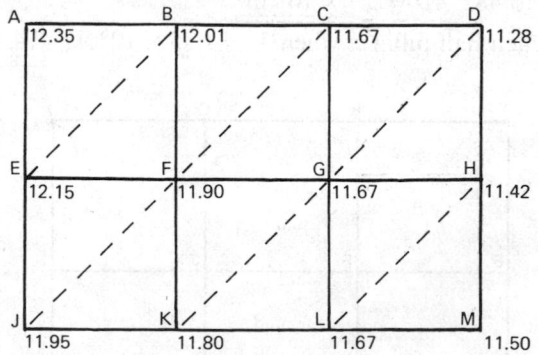

FIG. 66. *Specimen Question 30.*

SOLUTION

Pt.	Depth (m)	No. of times used, n.	nd
A	2.35	1	2.35
B	2.01	2	4.02
C	1.67	2	3.34
D	1.28	1	1.28
E	2.15	2	4.30
F	1.90	4	7.60
G	1.67	4	6.68
H	1.42	2	2.84
J	1.95	1	1.95
K	1.80	2	3.60
L	1.67	2	3.34
M	1.50	1	1.50
		24	42.80

$$\text{Volume} = 6 \times 15 \times 15 \times \frac{42.80}{24} = \underline{\underline{2407.5 \text{ m}^3}}$$

Clearly the accuracy will be improved by making the elements smaller because, each being approximated to a prism, the smaller they are the better the approximation. Hence, by dividing the plan into triangles as shown by the broken lines in Fig. 66, a more accurate result is obtained. It is left as an exercise for the reader to recalculate the volume using the triangles. Answer by this approach = 2396 m^3.

EMBANKMENTS AND CUTTINGS

In road and railway design, earthwork calculations are necessary at every embankment and cutting. Fig. 67 is a section through a typical cutting and shows the dimensions h, b, h_1, h_2, s_1 and s_2, which are generally the most convenient for calculating the area of the section. It should be noted that, turned upside down, so to speak, Fig. 67 represents a section through a typical embankment.

Briefly, the steps leading to the construction of a road are as follows.

(a) A preliminary survey establishes a likely route.

(b) A level run is carried out along the proposed centre-line, a series of levels being taken on lines at right-angles to the centre-line every 30 m.

(c) In the drawing office a longitudinal section along the centre-

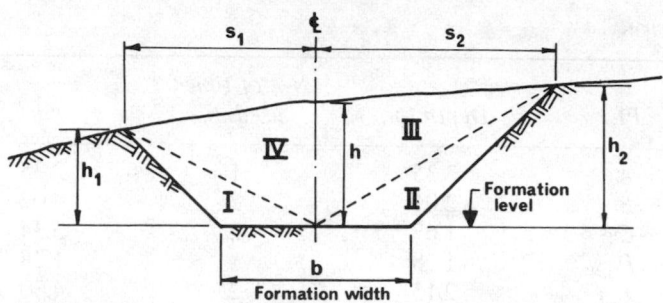

FIG. 67. *Section through a typical cutting.* $h = $ *centre-line height.* b = *formation width. The cross-sectional area is the sum of the areas of the triangles I, II, III, IV, and area* $= \frac{1}{2}b/2(h_1 + h_2) + \frac{1}{2}h(s_1 + s_2)$
$$= \tfrac{1}{4}b(h_1 + h_2) + \tfrac{1}{2}h(s_1 + s_2).$$

line is plotted and on it a trial elevation of the road is sketched, from which h can be determined.

(*d*) A suitable (depending on the density of traffic expected) forma-tion width b is chosen, side slopes compatible with the type of soil encountered are specified, and h is obtained from the longitudinal section. From these, cross-sections such as Fig. 67 can be plotted.

(*e*) The areas of the sections are calculated or measured by plani-meter, and from them the volume of each embankment or cutting is calculated by, for example, the prismoidal formula.

(*f*) A sixth stage is a mass haul diagram, which is essential for accurate determination of earthwork costs, *see* p. 146.

(*g*) Any or all of the above steps may have to be repeated once or twice or even more until it is thought that the most economic design has been obtained.

(*h*) When the drawing office work is satisfactorily completed the extremities of the side slopes are marked by pegs (usually set back about 1 m, to avoid being disturbed by the "muck-shifting" mach-inery), excavation or dumping is carried out to the desired formation level, and the road is constructed.

SPECIMEN QUESTION 41

A, B, C are three sections, 30 m apart, of a railway embankment, a typical cross-section being shown in Fig. 68 (*a*). The embankment is made on ground, the centre-line (X) of which is rising from A to C at a uniform slope of 1 in 20, the transverse slope remaining constant at 1 in 6.

The width of the bank at formation level is 18 m, and the side slopes are everywhere 1 in 3. The formation level also rises from A to C at a uniform slope of 1 in 60.

If the height of the embankment at the centre of the formation level at A is 6 m, find:

(a) the side widths, s_1 and s_2 at A, B and C;
(b) the cross-sectional areas at A, B and C;
(c) the volume of the embankment between A and C by the prismoidal formula.

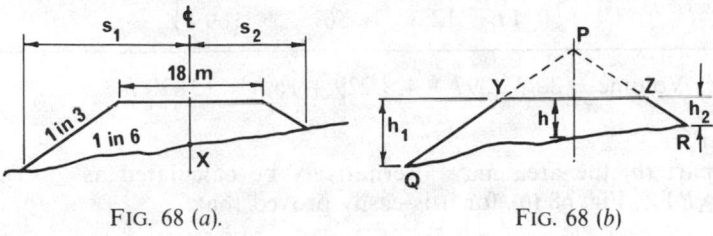

FIG. 68 (a).   FIG. 68 (b)

SOLUTION

(a) With h, h_1 and h_2 as shown in Fig. 68 (b),

$$s_1 - 9 = 3h_1 \qquad \text{(because side-slope} = 1/3),$$
$$h_1 = h + s_1/6 \qquad \text{(because cross fall of ground} = 1/6).$$

Eliminating h_1 between these two equations,

$$s_1 - 9 = 3(h + s_1/6)$$
$$\therefore s_1 - 3s_1/6 = 9 + 3h$$
$$s_1 = 2(3h + 9)$$

and $\therefore$ $\qquad h_1 = h + s_1/6,$
i.e. $\qquad h_1 = 2h + 3$

Similarly, $\qquad s_2 - 9 = 3h_2$
and $\qquad h_2 = h - s_2/6$
$$\therefore s_2 - 9 = 3(h - s_2/6)$$
$$\therefore s_2 = 2/3 \,.\, (3h + 9)$$
and $\qquad h_2 = 2/3 \,.\, h - 1$

Since the ground at X rises at 1 in 20, i.e. by 1.5 m in every 30 m and, since the formation level rises at 1 in 60, i.e. by 0.5 m in every 30 m, h decreases by 1 metre in every 30 metres. Hence:

Section	h	$3h + 9$	s_1	s_2	
A	6	27	54	18	
B	5	24	48	16	*Answer* (a)
C	4	21	42	14	

(b) From the formula quoted in the caption to Fig. 67

$\qquad$ Area of section $= \frac{1}{4}b(h_1 + h_2) + \frac{1}{2}h(s_1 + s_2)$
i.e. $\quad$,, $\quad$,, $\quad$,, $= 4.5(h_1 + h_2) + \frac{1}{2}h(s_1 + s_2)$. Hence:

Section	h	h_1	h_2	$h_1 + h_2$	$s_1 + s_2$	Area (m^2)
A	6	15	3	18	72	297
B	5	13	2.$\dot{3}$	15.$\dot{3}$	64	229
C	4	11	1.$\dot{6}$	12.$\dot{6}$	56	169

$\left. \begin{matrix} \\ \\ \end{matrix} \right\}$ *Answer* (b)

(c) Volume = $30/3 . (297 + 4 \times 229 + 169) = 13\,820$ m^3

NOTE

In part (b) the area may alternatively be calculated as $\triangle PQR - \triangle PYZ$, Fig. 68 (b), for it is easily proved that:

$$\triangle PQR = \tfrac{1}{2}(h + 3)(s_1 + s_2), \text{ and}$$
$$\triangle PYZ = \tfrac{1}{2} \times 18 \times 3 = 27 \text{ m}^3.$$

SPECIMEN QUESTION 42

A road having a formation width of 40 m with side slopes of 1 to 1 is to be constructed. Details of two cross-sections of a cutting are as follows:

Chainage	Depth of cutting on centre line (m)	Side slope limits (m) Left	Right
500	10.2	25.2	33.7
530	6.0	22.0	28.5

Assuming that these cross-sections are bounded by straight lines and that the undisturbed ground varies uniformly between them, compute the volume of excavation allowing for prismoidal excess.

If, instead of being straight, the plan of the centre-line had been a circular curve of radius R with the centre of curvature on the right, how would this have been taken into account in the foregoing calculations? Quote any formula that would have been used.

SOLUTION

Although other means are at hand, the simplest way of allowing for prismoidal excess is probably to use the prismoidal formula, which, in this question, necessitates finding the dimensions of a section midway between the given sections.

To do this it is usual to make the assumption that corresponding points on the two end sections are joined by straight lines. The dimensions of the middle section are then given by the mean of the corresponding dimensions on the end sections.

e.g. depth on centre-line, $h = \frac{1}{2}(10.2 + 6.0) = 8.1$

In Fig. 69 (a) the dimension 5.2 m is derived from the side slope limit of 25.2 m and the fact that the side slopes at 1 in 1. The dimensions 13.7 m, 2 m and 8.5 m are similarly obtained.

At chainage 515 m
$$s_1 \tfrac{1}{2}(25.2 + 22.0) = 23.6 \text{ m}$$
$$s_2 = \tfrac{1}{2}(33.7 + 28.5) = 31.1 \text{ m}$$
$$\therefore h_1 = 3.6 \text{ m}$$
and
$$h_2 = 11.1 \text{ m}$$

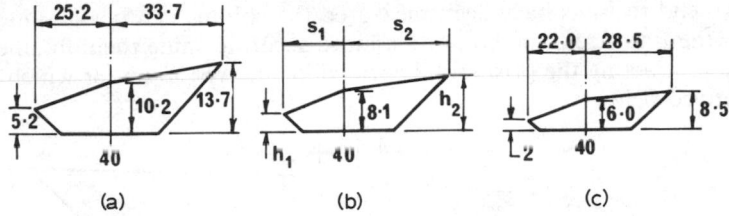

FIG. 69. *Cross-sections in Specimen Question 42; (a) at chainage 500 m, (b) at chainage 515 m, and (c) at chainage 530 m*

The formula given in Fig. 67 is used to calculate the areas. These are:

At chainage 500:

$$\text{Area} = \tfrac{1}{2} \times 20(5.2 + 13.7) + \tfrac{1}{2} \times 10.2(25.2 + 33.7) = 490 \text{ m}^2$$

At chainage 515:

$$\text{Area} = \tfrac{1}{2} \times 20(3.6 + 11.1) + \tfrac{1}{2} \times 8.1(23.6 + 31.1) = 368 \text{ m}^2$$

At chainage 530:

$$\text{Area} = \tfrac{1}{2} \times 20(2.0 + 8.5) + \tfrac{1}{2} \times 6.0(22.0 + 28.5) = 257 \text{ m}^2$$
$$\therefore \text{Volume} = 15/3 \,.\, (490 + 4 \times 368 + 257) = 11\,100 \text{ m}^3$$

Correction for curved centre-line. The centroid of the cross-section at chainage 500 m obviously lies to the right of the centre-line, i.e. on the same side of the centre-line as the centre of curvature. Suppose it is a distance c from the centre-line; then, to allow for the curvature of the centre-line, the area of this section must be taken as $490 - 490c/R$ instead of 490 m². If the centroid lay on the side of the centre-line remote from the centre of curvature the correction $490c/R$ would have to be added.

The volume would be calculated by making a similar correction to the area of each of the three sections and substituting the revised areas in the prismoidal formula, still using the factor 15/3.

Prismoidal correction
This is defined by:

prismoidal correction = Volume by End Areas
− Volume by Prismoidal formula

and has a particular relevance to calculations of the volumes of cuttings and embankments. Provided that it is bounded by straight lines a cutting or embankment conforms exactly to the surveyor's definition of a prismoid and its volume can be calculated very accurately (exactly, in theory) by the prismoidal formula. If only the two end sections have been measured the volume can be computed by the End Areas method and a more accurate value then obtained by subtracting the prismoidal correction, an expression for which is derived below.

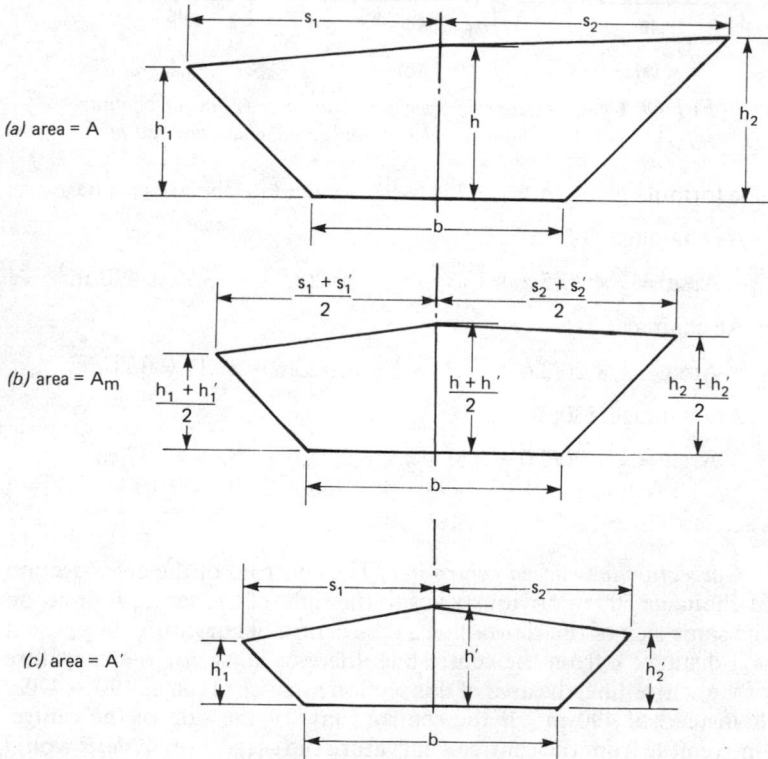

FIG. 70. *The dimensions of the middle section (b) of a cutting, given the dimensions of the end sections (a) and (c).*

The volume can be calculated by the prismoidal formula if, first, the middle area is calculated. The formation width, b, is the same at every section but all the other dimensions in the middle section are the means of the corresponding dimensions in the end sections of area A and A' respectively, see Fig. 70.

Let middle area $= A_m$ and note that A_m is not the mean of A and A'.

Let $d =$ distance between the two end sections. Then:

$$\text{by prismoidal rule, volume} = \tfrac{1}{3} \times \frac{d}{2}(A + 4A_m + A')$$

$$\text{by end areas, volume} = \tfrac{1}{2} \times d(A + A')$$

$$\therefore \text{prismoidal correction} = \left(\frac{d}{2} - \frac{d}{6}\right)(A + A') - \frac{d}{6} \times 4A_m$$

$$= \frac{d}{3}(A + A' - 2A_m)$$

Now,

$$A = \frac{b}{4}(h_1 + h_2) + \frac{h}{2}(s_1 + s_2)$$

$$A' = \frac{b}{4}(h_1' + h_2') + \frac{h}{2}(s_1' + s_2')$$

$$A_m = \frac{b}{4}\left(\frac{h_1 + h_1'}{2} + \frac{h_2 + h_2'}{2}\right) + \frac{h + h'}{4}\left(\frac{s_1 + s_1'}{2} + \frac{s_2 + s_2'}{2}\right)$$

$$\therefore A + A' = \frac{b}{4}(h_1 + h_2 + h_1' + h_2') + \frac{h}{2}(s_1 + s_2) + \frac{h'}{2}(s_1' + s_2') \quad \text{and}$$

$$2A_m = \frac{b}{4}(h_1 + h_2 + h_1' + h_2') + \frac{h + h'}{4}(s_1 + s_2 + s_1' + s_2')$$

$$\therefore A + A' - 2A_m = \frac{h}{2}(s_1 + s_2) + \frac{h'}{2}(s_1' + s_2') - \frac{h + h'}{4}(s_1 + s_2 + s_1' + s_2')$$

$$= (s_1 + s_2)\left(\frac{h}{4} - \frac{h'}{4}\right) - (s_1' + s_2')\left(\frac{h}{4} - \frac{h'}{4}\right)$$

$$\therefore \text{prismoidal correction} = \frac{d}{3}\left(\frac{h - h'}{4}\right)\left((s_1 + s_2) - (s_1' + s_2')\right)$$

The above bookwork shows that for cutting calculations, and therefore embankment calculations too, the prismoidal correction is exactly the difference in volumes as calculated by end areas and by the prismoidal formula. The reader should now work through

Specimen Question 42 using end areas and applying the prismoidal correction to get exactly the same answer.

MASS HAUL DIAGRAM

This chapter has dealt so far with the various methods of calculating areas and volumes. In building and civil engineering practice the calculation must go further and lead to an estimation of the cost involved in carrying out the earthworks. The best solution will not necessarily be the one requiring the least excavation but, rather, the one which minimises the cost while meeting all the other requirements of the construction. And this is where the Mass Haul Diagram is used—in comparing costs of a number of trial solutions.

The basis of how a mass haul diagram is constructed is illustrated by the next examination question.

SPECIMEN QUESTION 43
The following table gives the area of cut and fill necessary at intervals along the line of a proposed new road as determined from measured cross sections.

Chainage	Cut (m^2)	Fill (m^2)
0	—	—
50	22	—
100	47	—
150	46	—
200	28	—
250	12	—
300	1	—
350	—	3
400	—	34
450	—	57
500	—	52
550	—	48
600	—	21
623	—	7

(a) Calculate the total volume of cut.

(b) If all the excavated material is to be used as fill, and the distance of transporting this material is to be kept to a minimum, calculate the chainage at which all the excavated material will have been used.

(c) What volume of fill material will have to be imported to the site in order to complete the earthworks?

SOLUTION

First, the relevant volumes are calculated thus:

Chainage	areas (m^2) Cut	Fill	volumes (m^3) between sections Cut	Fill	Accumulated Volume (m^3)
0	—	—	—	—	—
50	22	—	550	—	550
100	47	—	1725	—	2275
150	46	—	2325	—	4600
200	28	—	1850	—	6450
250	12	—	1000	—	7450
300	1	—	325	—	7775
350	—	3	—	50	7725
400	—	34	—	925	6800
450	—	57	—	2275	4525
500	—	52	—	2725	1800
550	—	48	—	2500	− 700
600	—	21	—	1725	− 2425
623	—	7	—	322	− 2747

In the above table cuts, i.e. excavations, are considered to be positive and fills negative. The volume (m^3) between sections are calculated by the end areas method applied successively between adjacent sections thus:

$$550 = \frac{50}{2}(0 + 22)$$

$$1725 = \frac{50}{2}(22 + 47)$$

The accumulated volume is the total of *cut* measured from chainage 0. Thus:

at chainage 250 accumulated volume = 7450 m^3

„ „ 400 „ „ = 6800 m^3

(= net cut − fill)

„ „ 550 „ „ = − 700

(i.e. fill exceeds cut)

It follows that the total volume of excavation is given by the maximum value of the curve, *see* Fig. 71 and is 7800 m^3 (*a*)

It also follows that *cut* and *fill* are equal when the curve crosses the x-axis, *see* Fig. 71, i.e. at chainage 535 m(*b*)

The last ordinate on the graph, namely -2747 m³ shows that the fill requirement exceeds the volume available from the cutting by this amount. i.e.

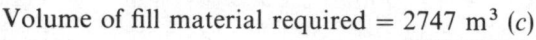

Volume of fill material required = 2747 m³ (*c*)

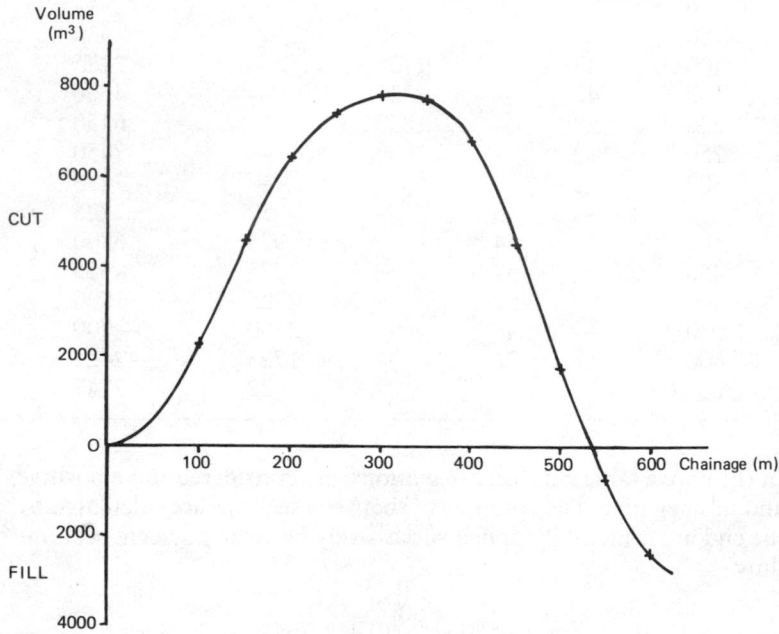

FIG. 71. *See Specimen Question 34.*

In the above example the graph is a Mass Haul Diagram which is the name given to a plot of accumulated volume of excavation against chainage. In earthworks calculations this curve is often presented with a longitudinal section through the proposed work (say a major road), each curve being in strict projection with the other, *see* Fig. 73.

The main features of the mass haul diagram are:

(*a*) positive slope corresponds to *cut*;
(*b*) negative slope corresponds to *fill*;
(*c*) the ordinate (i.e. *y* co-ordinate) at any chainage gives the net value of volume of cut − volume of fill.

When the curve is below the x-axis, therefore, fill exceeds cut;
(*d*) When the curve crosses the x-axis, i.e. at *y* = 0, volume of

cut = volume of fill, i.e. cut and fill balance and, therefore, the x-axis is called a balancing line.

In fact, any line which is parallel to the x-axis and which also cuts off a loop of the mass haul diagram is a balancing line because, as a few moments of thought will show, cut and fill will balance along such a line. The area under the curve is a measure of volume × distance carried and this product is called Haul.

A simple application of the mass haul diagram would be to calculate the average transportation distance. In Specimen Question 43, for example, the area under the loop is approximately 2.5×10^6 m^4. Since the volume being transported is 7800 m^3, the average distance each truck load has to be carried is $2.5 \times 10^6 \div 7800 = 320$ m approximately. Furthermore, if the trucks are capable of carrying 10 m^3 each, the number of truck journeys would be 780. If only one truck were used it would have to travel 780×0.32 km = 250 km each way — and this kind of calculation would help the engineer to take decisions on the type and number of trucks required.

A more complete example on Mass Haul Diagrams

Specimen Question 43 and the notes associated with it has served as a simplified introduction to Mass Haul Diagrams (i.e. M.H.D.)—a halfway step, as it were, towards a fuller comprehension. Below, therefore, a more complete calculation is performed and it should be noted how it follows the steps outlined on p. 139. From the survey a longitudinal section has been plotted along the route of the proposed road and a trial formation level centre-line drawn on it, *see* Fig. 73 (*a*). The height h has been scaled off at intervals of 200 m. To simplify the arithmetic it has been assumed that every cross-section is symmetrical about the centre-line, as shown in Fig. 72, and that with the main dimensions as shown,

$$\text{cross-sectional area, csa} = 30\ h + \tfrac{3}{4}\ h^2$$

Hence the csa can be calculated at every 200 m and tabulated.

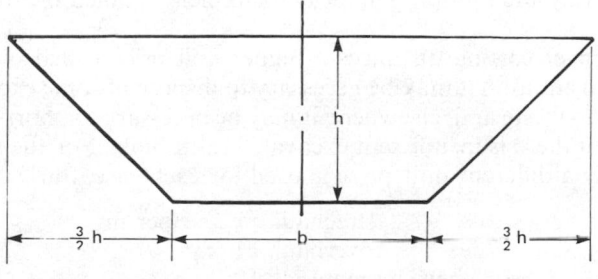

FIG. 72

The volume between adjacent cross-sections has then been calculated by the end areas method, cuts being shown as positive and fills as negative. The table was then completed by calculating the "Accumulated Volume" which is, in fact, the running total of the figures in the previous column.

Whence:

Chainage	"h"	csa (m^2)	volume (m^3)	Accumulated volume (m^3)
0	0	0	0	0
200	2.0	63	6300	6300
400	3.6	118	18 100	24 400
600	2.4	76	19 400	43 800
800	-2.4	-76	0	43 800
1000	-5.6	-192	$-26 800$	17 000
1200	-5.6	-192	$-38 400$	$-21 400$
1400	-1.2	-37	$-22 900$	$-44 300$
1600	2.8	90	5300	$-39 000$
1800	5.2	176	26 600	$-12 400$
2000	4.8	161	33 700	21 300
2200	-1.6	-50	11 100	32 400
2400	-4.4	-146	$-19 600$	12 800
2600	-0.5	-15	$-16 100$	-3300

The graphs shown in Fig. 73 were then plotted.

The aim of this exercise is to consider various ways of moving excavated material to form embankment and to try to find out which is cheapest. Since it obviously costs more to cart fill material over long distances than it does over short distances, an arbitrary distinction based on experience has been made between short hauls and long hauls, the dividing line being taken as 500 m. This is sometimes called the free haul distance and the unit price, which includes excavating, carting away and dumping in the embankment is called the free haul price.

For longer carting distances a higher unit price called overhaul is used. In addition it may be necessary to dispose of some excavated material to waste and, elsewhere, it may be necessary to "borrow" fill material if there is insufficient excavated in the cuts. For the sake of generality a different unit price is used for each case, thus:

$$\begin{array}{lll} \text{freehaul:} & \text{£}x & \text{per m}^3 \\ \text{overhaul:} & \text{£}1.4x & \text{,,} \quad \text{,,} \\ \text{waste material:} & \text{£}2x & \text{,,} \quad \text{,,} \\ \text{borrowed material:} & \text{£}2.5x & \text{,,} \quad \text{,,} \end{array}$$

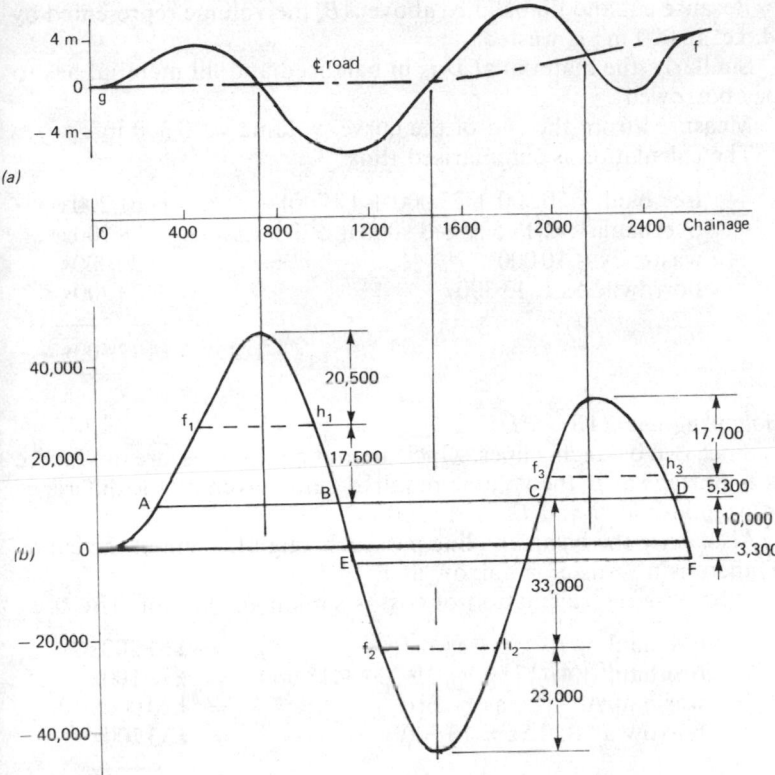

FIG. 73. (a) *Longitudinal section.* (b) *Mass Haul Diagram.*

On Fig. 73 (b), the horizontal lines f_1h_1, f_2h_2, f_3h_3 represent the free haul distance of 500 m. They are obtained by finding where a horizontal length of 500 m to scale intercepts each of the loops.

The volumes cut off by the "*fh*" lines is then read off the vertical axis and, for clarity, is marked on the graph: 20 500, 23 000 and 17 700 m³ respectively. These are the volumes at the free haul price.

A number of trial balancing lines are now drawn and the cost calculated for each.

Balancing Line ABCD
Remember that in the loop cut off by the balancing line cut and fill balance. Two prices apply above AB: the free haul price above f_1h_1 and the overhaul price for the rest, i.e. 17 500 m³.

Similarly, below BC there is 23 000 m³ at £x and 33 000 at £1.4x

Similarly, above CD there is 17 700 m³ at £x and 5 300 at £1.4x

Because cut and fill balance above AB, the volume represented by A, i.e. 10 000 m³ is wasted.

Similarly, the material at D is unbalanced and fill material has to be "borrowed".

Measured from the end of the curve, volume = 13 300 m³.

The calculation is summarised thus:

free haul: x(20 500 + 23 000 + 17 700)	= £	61 200x
overhaul: 1.4x(17 500 + 33 000 + 5300)	=	78 100x
waste: 2x × 10 000	=	20 000x
borrow: 2.5x × 13 300	=	33 200x
	Total =	£192 500x

Balancing lines $AB + EF$

Since two balancing lines, which are not continuous, are used there is a shortage at B, the volume required being given by the difference in ordinates at B and E, i.e. 13 300 m³.

Also, since the balancing line passes through the end of the curve, F, there is no waste or borrow at F.

Otherwise the calculation of cost is similar to that for $ABCD$:

free haul: as before	= £61 200x
overhaul: 1.4x(17 500 + 19 700 + 18 600	= £78 100x
waste at A: as before	= £20 000x
borrow at B: 2.5x × 13 300	= £33 200x
Total =	£192 500x

Balancing line along the x-axis

free haul: as before	= £61 200x
overhaul: 1.4x(27 500 + 23 000 + 15 300)	= £92 100x
borrow: 2.5x × 3300	= £8200x
Total =	£161 500x

Clearly, the x-axis is the best of these three balancing lines to use. The practical significance of this decision is that excavation should be commenced at chainage 0 and tipped where the ground level falls below proposed road level, i.e. at chainage 700 + approximately. Excavation should continue between chainages 0 and 700 until the cutting is finished and the embankment completed as far as chainage 1100 approximately. The rest of the fill material required between chainages 1100 and 1460 is obtained from the excavation beyond chainage 1460. A relatively small amount of borrowed fill material is required between chainages 1580 and 1600.

Every balancing line calculation lends itself to this kind of interpretation and the reader should, therefore, practice by interpreting what the other two balance lines mean in terms of earth movement.

It should be noted that in practice, calculations such as the above have to be modified slightly to allow for the bulking and shrinkage which accompany the excavation and subsequent replacement and compaction of the material. This can be taken into account by multiplying the figures in the "Volume" column by the appropriate factor before compounding them to give the "Accumulated Volume".

The next stage would be to try another trial formation level, recalculate h, areas and volumes, draw another MHD and calculate the costs with two or three trial balancing lines—*see* Examination Question 1 below.

EXAMINATION QUESTIONS

1. Re-draw to scale the ground profile shown in Fig. 73. Draw on it a new trial formation level for the road:

(a) horizontal through g;
(b) rising with uniform gradient from g to f

and calculate the cost in each case for a balancing line along the x-axis.

2. A small parcel of land has been surveyed so that the area may be calculated. A and B are two points on the fence surrounding the area. A is the southern point and B the northern one. A survey line running north and south across the area was measured between the points A and B and offsets taken at right angles to the survey line to each change of direction in the boundary. These measurements are given below, in metres:

West	Chainage	East
0	300	40
	275	45
30	250	34
24	220	
	200	26
10	150	38
	125	40
37	60	
	50	20
0	0	10

Calculate without plotting, the area in hectares contained in the parcel of land.

3. (*a*) Calculate the amount of sandstone, in cubic metres, in a triangular area, the sides of which are 451.20 m, 321.70 m and 208.90 m. The sandstone has a thickness of 2.00 m and lies horizontally.

(*b*) The base of a heap of debris deposited on level ground is a 400 m × 400 m square and the top is a 120 m × 120 m square. The heap is symmetrical and is 30 m high. Calculate the volume of the debris in cubic metres.

4. Figure 74 shows (not to scale) the cross section of an earth embankment. The existing ground on which the embankment is to be made has a gradient of 1 (vertical) in 7 (horizontal) and the side slopes are to be 1 in 2.

The top width is to be 10 m and the depth on the centre line 4 m. Determine the distances *OA* and *OB* measured along the slope from the centre line to the limits of the embankment.

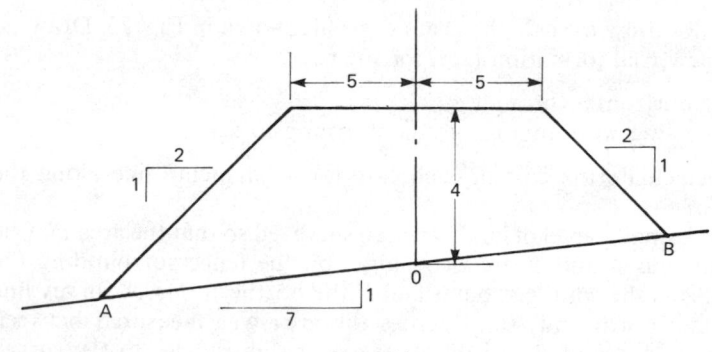

FIG. 74

5. (*a*) Figure 75 (*a*) shows contours on the bed of a lake and the table gives the total plan area within each contour.

Estimate as accurately as you can the volume of water in the lake when the water surface is at 108 m A.O.D.

Contour level m A.O.D.	Plan area m^2
100	200
102	800
104	4 000
106	8 000
108	11 000

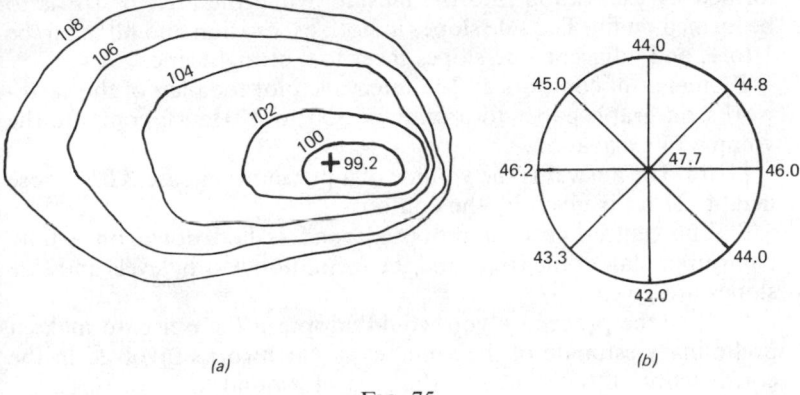

(a) (b)

FIG. 75

(b) Figure 75 (b) shows the plan of a proposed vertical-sided excavation with existing ground levels at eight symmetrical positions around the circumference and at the centre. If the excavation is 20 m in diameter and is to have a flat bottom at a reduced level of 40.0 m, estimate the volume of material to be excavated.

6. A storage tank is to be built in a hillside for irrigation purposes. The formation level of the foundation of the tank is to be 2000 m A.O.D., the sides are to be vertical and the outside measurements 90 m by 45 m. The existing ground surface was gridded and levelled at 15-m intervals, and a table of the recorded spot heights is given below. Compute the volume of excavation required in m^3.

2034.8	2031.1	2028.7	2025.2	2026.6	2029.3	2032.5
2029.4	2027.6	2024.3	2022.7	2023.2	2025.8	2027.9
2024.1	2023.9	2020.2	2019.8	2020.3	2021.7	2022.6
2020.5	2018.4	2016.6	2015.3	2015.7	2016.2	2017.8

7. The areas of cross-sections along the centre-line of a proposed embankment are as follows (measurement in metres):

Distance	0	30	60	90	120	150	180	210	240
Cross sectional areas (m^2) *above datum*	89.24	90.36	90.25	91.13	91.74	90.58	90.12	89.47	90.68

Calculate the volume of material in the embankment by both the prismoidal and trapezoidal rules.

8. A square level area *ABCD* (in clockwise order) of 100-m side is to be formed in a hillside which is considered to have a plane surface with a maximum gradient of 3 horizontally to 1 vertically. *E* is a point which bisects the side *AD*, and the area *ABE* is to be

formed by excavation into the hillside, while the area $BCDE$ is to be formed on fill. The side slopes in both excavation and fill are to be 1 to 1, and adjacent side slopes meet in a straight line.

By means of contours at 2-m intervals, plot the plan of the earthworks on graph paper to a scale of 500 to 1. Hence compute the volume of excavation.

[*Hint.* The answer is the volume of a pyramid on base ABE whose height (27 m) is given by the contours.]

9. The centre-line of a proposed road is laid down on a fully contoured plan of the route and the formation width, levels and side slopes are given.

Outline the procedure you would adopt, in the office, to make a preliminary estimate of the volumes of earthworks involved in the construction of the road and the area of ground to be disturbed.

10. A section of a proposed road is to run through a cutting from chainage 500 to 900, the formation level falling at 1 in 200 from chainage 500. The formation width is to be 30 m and the side slopes are to be 1 vertical to 2 horizontal. The original ground surface is inclined uniformly at right-angles to the centre-line at an inclination of 1 in 10.

With the information given below, calculate the volume of excavation in cubic metres, using the prismoidal formula.

Chainage	Formation level	Ground level at centre-line
500	44.25 m	51.11
600		50.82
700		50.93
800		51.09
900		50.77

11. The centre-line of a section of a proposed road in cutting is indicated by pegs at equal intervals, and the corresponding longitudinal section gives the existing ground level and the proposed formation level at each peg, but no cross-sections have been taken, or sidelong slopes observed.

Given the proposed formation width (d) and the batter of the sides (S horizontal to 1 vertical), how would you set out the batter pegs marking the tops of the slopes at each centre-line peg, without taking and plotting the usual cross-section.

An alternative method would be acceptable.

CHAPTER 10

FIELD ASTRONOMY

The Star Almanac for Land Surveyors, published annually by H.M. Stationery Office, is an essential book for surveyors using the methods of field astronomy. The surveying student must be familiar with its contents; it is often referred to in this chapter, and it is assumed that the reader will have bought or borrowed a copy.

THE OBJECT OF FIELD ASTRONOMY

Usually the object of the surveyor is to establish a direction accurately; in addition, he may, possibly as a preliminary, have to determine the latitude of his position.

Figure 76 shows some of the main features of the celestial sphere, an artifice of infinite dimensions upon which all the heavenly bodies are supposed to lie, and which is considered to revolve about the axis PP' roughly once every day, the earth being considered stationary.

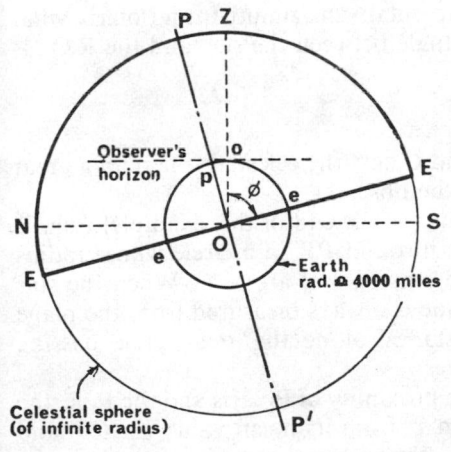

p is the north pole
ee is the equator
o the observer's position
Z is the observer's *zenith*, the point in the heavens directly above him
P is the celestial north pole
P' is the celestial south pole
EE is the celestial equator
O is the centre of the celestial sphere
ϕ is the observer's latitude
NS may be taken as the observer's horizon because, the celestial sphere being infinite, $oO/ON =$ zero

FIG. 76. *The celestial sphere for an observer in the northern hemisphere.*

DETERMINING AZIMUTH

Because distances to the stars are so great (the nearest is about four light years away), it is permissible to take o and O as the same point. In Fig. 76, if a theodolite could be set up at o and sighted on to P the telescope would be inclined to the horizontal at angle $PON = \phi$, and the vertical scale reading would therefore give the latitude of the instrument station. If, further, the horizontal scale was read with the telescope pointing to P (i.e. due north), and again when pointing to a reference object (R.O.), the difference between these readings would be the azimuth* of R.O., thus establishing its direction. In short, by pointing the theodolite telescope to P, and again to the R.O., the surveyor would achieve his objects.

Sighting Polaris
Unfortunately there is no star at P, which is therefore a point in the heavens which cannot be identified. A method based on sighting Polaris, the star nearest the celestial north pole, does exist, however, and is described in Specimen Question 44. It is a modification of the routine just described.

SPECIMEN QUESTION 44
Draw a diagram of the visible portion of the celestial sphere for an observer in latitude 52° N and indicate on the diagram the diurnal path of Polaris. Explain why Polaris is a suitable star for the determination of azimuth.

If Pole Star tables are used to obtain an azimuth from Polaris, what information, other than the angle between the star and the R.O., is required and in what form?

SOLUTION
The plane containing P, Z and O cuts the celestial sphere in a great circle called the meridian of the observer.

In one sidereal day (*see* notes at the end of this example) Polaris makes one complete revolution round PP' in a circle whose radius subtends an angle of about 54 minutes of arc at O. When the star is at X its altitude is α. Altitude is always measured from the plane of the horizon, NS, to the star X along the great circle passing through Z and X.

In Fig. 77 (*a*), because of the proximity of Polaris and the Pole, the angle ZOX will vary less than 1° from its mean value $\angle ZOP$, and consequently the altitude will vary by less than 1° from the latitude.

---

* The azimuth of B from A is the horizontal angle, measured clockwise between AB and true north at A. It is not the same as the azimuth of A from B, which, of course, is measured from true north at B.

Fig. 77 (*b*) shows that, providing *Z* is not too near *P*, the angle *PZX* is always small and therefore changes very slowly as the star moves round its daily path. It follows that the time at which observation is made (which gives the instantaneous position of the star) is not so critical as it would otherwise be, for a small error in time corresponds to a very much smaller error in the azimuth angle (*PZX*). Repeating this statement in the language of field astronomy, we say: provided the observer's latitude is not too great (the *Star Almanac* includes no latitudes greater than 66° N) Polaris moves very slowly in azimuth, minimising any error in the recorded time of observation.

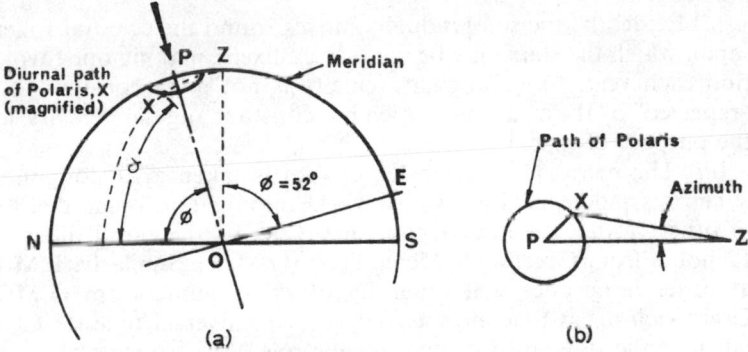

FIG. 77. (*a*) *The visible portion of the celestial sphere for an observer in latitude 52° N.* (*b*) *View of* (*a*) *from the arrow* (*to an enlarged scale*).

The reasons why Polaris is a suitable star from the determination of azimuth are:

(*a*) Its slow movement in azimuth.

(*b*) It is a moderately bright star.

(*c*) It is easily identified, and

(*d*) Pole Star tables are available which make the calculation of azimuth (and latitude) very simple.

The Pole Star tables are based on two formulae (*see* the *Star Almanac*), which are:

$$\text{Latitude, } \phi = \alpha + a_0 + a_1 + a_2$$

where α = corrected observed altitude of Polaris (*see* p. 167);

$$\text{Azimuth} = (b_0 + b_1 + b_2) \sec \phi$$

a_0, a_1, a_2, b_0, b_1 and b_2 are six parameters which depend on the time of year, the approximate latitude and the local sidereal time (L.S.T.) at which the observation is made. Hence, to determine azimuth, these quantities must be known; in addition, the observer's longitude must be known (for the L.S.T. depends on longitude), and barometric pressure and temperature must be known for calculating the refraction correction (*see* the refraction tables in the *Star Almanac*) to the observed altitude.

NOTES

(*i*) Considering the earth as stationary, in a year the sun makes approximately $365\frac{1}{4}$ revolutions and the stars $366\frac{1}{4}$ revolutions around it. Thus, the time for a star to make exactly one revolution, namely one sidereal day, is $365\frac{1}{4}/366\frac{1}{4}$ of a day by the sun. In other words, a sidereal day is about 4 minutes shorter than a day by the mean sun.

(*ii*) Evidently, the sun gradually moves round the celestial sphere (upon which the stars may be considered fixed), making one revolution each year. As its angular velocity is not quite constant, it is "replaced" by the mean sun, which has constant angular velocity, for the purpose of time.

(*iii*) The meridian through Greenwich is taken as $0°$ longitude. When it is midday in Greenwich it will be midnight on longitude $180°$; in other words, the local mean time (L.M.T.) there will differ by 12 hours from Greenwich Mean Time (G.M.T.). Similarly, L.M.T. at other longitudes will differ by other amounts from G.M.T. Greenwich mean time, now usually called universal time (U.T.), is taken as the standard for time calculations.

It follows that one hour of time is equivalent to $15°$ of arc; in fact, angles in field astronomy are often measured in hours, minutes and seconds, 1 second of time being 15 seconds of arc.

(*iv*) The difference between sidereal and mean time (the 4 minutes per day) gradually accumulates during the year, the two times being the same only once per year, about September 22nd. The difference, R, is tabulated in the *Star Almanac*.

The L.S.T. anywhere, at any time, can be calculated from:

$$\text{L.S.T.} = R + \text{U.T.} \left. \begin{array}{l} - \text{ west} \\ + \text{ east} \end{array} \right\} \text{longitude}$$

(*v*) Only planes which pass through its centre cut a sphere in a *great circle*. It follows that one, and only one, great circle passes through any two points on the surface of a given sphere, provided the points are not diametrically opposite one another.

The celestial triangle

It may not be convenient to use Pole Star tables, or the results they give may not, in some circumstances, be sufficiently accurate. A number of alternative methods exist, nearly all of which depend on solving a particular spherical triangle variously known as the celestial triangle, the astronomical triangle or the navigational triangle. A description of this spherical triangle forms the basis of Specimen Question 45.

By definition the sides of a spherical triangle are arcs of great circles; the lengths of the sides are usually described in degrees, minutes and seconds of arc, i.e. by the angles they subtend at the centre of the sphere.

SPECIMEN QUESTION 45

Describe with clear diagrams the component parts of the celestial triangle and indicate two ways of solving the triangle to give a value for the azimuth angle. Also draw the celestial triangle, indicating its side length, for the special case of a circum-polar star at elongation. For what purpose is a star at elongation observed and what are the particular merits of such an observation?

SOLUTION

Figure 78 (*a*) is a sketch of the celestial sphere viewed from the east. The same lettering has been used as in previous figures, except that X is any star, the sun or a planet. The spherical triangle PZX is the celestial triangle; an enlarged view is given in Fig. 78 (*b*).

The arc, from the plane of the equator to X, measured along the great circle through P and X, is defined as the declination of X. Hence, bearing in mind the definitions already given of latitude and altitude, the sides of the celestial triangle (Fig. 78 (*b*)) are:

$$x = 90° - \phi = \text{co-latitude}$$
$$z = 90° - \delta = \text{co-declination} = \text{polar distance}$$
$$p = 90° - \alpha = \text{co-altitude} = \text{zenith distance}$$

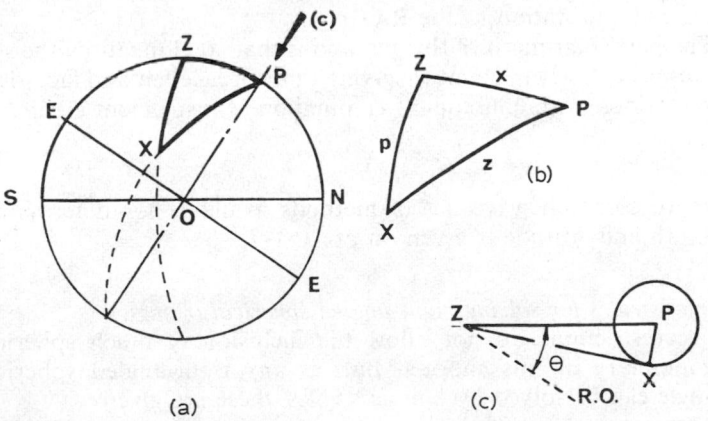

FIG. 78. *The celestial triangle* ZPX

The angles of the celestial triangle are:

$$Z = \text{azimuth angle}$$
$$P = \text{hour angle}$$
and
$$X = \text{parallactic angle}$$

To solve the celestial triangle it is necessary to know any three of the above listed six parts. As the path of a star is roughly parallel to the celestial equator—the *Star Almanac* shows that declination varies a few seconds from month to month, contrary to our earlier assumptions about "fixed stars"—its declination is known, and hence z is known. If the observer's latitude is known, x is known, and the spherical triangle can be solved by measuring either the altitude, to give p, or the hour angle, to give P. Altitude is measured direct by theodolite; to determine hour angle, the L.S.T. at the instant of observation must be calculated.

Elongation. Every star crosses the meridian twice: at upper culmination, the crossing nearest to Z, and again at lower culmination. Referring to Fig. 78 (*a*), a circum-polar star is one which culminates between Z and P, and again between P and N. Fig. 78 (*c*) is a view of the celestial sphere in the direction of the arrow in Fig. 78 (*a*). In Fig. 78 (*c*), P and Z being fixed points, the circum-polar star X appears to be farthest from P when the angle X is $90°$. The star is then said to be at east (as shown) or west elongation.

The purpose of observing a star at elongation is again to determine an azimuth. In Fig. 78 (*c*), for example, if z and x are known the triangle can be solved (*see* Specimen Question 46). Thus, if the horizontal scale of a theodolite is read first with the telescope pointing at the star at elongation, and secondly, at a reference object (R.O.), the angle θ can be determined. θ is the azimuth of the line from the theodolite station to the R.O.

The particular merit of this method is that, at elongation, the star moves very slowly in azimuth, giving time for face-left and face-right observations, with consequent elimination of instrument errors.

NOTE

A more comprehensive list of methods available for determining azimuth and latitude is given on pp. 164–7.

Napier's rules for solving right-angled spherical triangles
Space restrictions do not allow the inclusion of much spherical trigonometry in this chapter, but, as any right-angled spherical triangle can be solved by Napier's rules, these are given.

Any sector in Fig. 79 (*b*) can be taken as the middle part. Then the rules state:

(*a*) Sin (middle part) = Product of cosines of opposite parts, e.g.

$$\sin (\pi/2 - C) = \cos c \cdot \cos (\pi/2 - A)$$
$$\therefore \cos C = \cos c \cdot \sin A$$

(*b*) Sin (middle part) = Product of tangents of adjacent parts, e.g.

$$\sin a = \tan (\pi/2 - C) \cdot \tan c$$
$$\therefore \sin a = \cot C \cdot \tan c$$

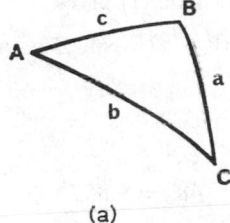 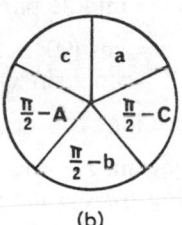

(a) (b)

FIG. 79. *Napier's rules. (a) ABC is a spherical triangle with a right-angle at B. (b) The five parts are the sides enclosing the right-angle, the complement of the other side, and the complement of the two angles other than the right-angle. Notice that the five parts occur in the same cyclic order on both figures.*

SPECIMEN QUESTION 46

With the aid of a sketch, explain how to determine whether a particular star will elongate when viewed from a given latitude.

A star of declination 61° 57′ 24″ N is observed at eastern elongation from a point *A*, which is at latitude 43° 58′ 12″ N. At elongation the horizontal clockwise angle from the star to a point *B* is 83° 12′ 12″. Compute the azimuth of *B* from *A*.

SOLUTION

The explanation for the first part of the question follows from Specimen Question 45. It is left to the reader to show that, in terms of the declination δ and the latitude ϕ, the star will elongate only if $\delta > \phi$. Such a star will not necessarily be circumpolar; its lower transit will be above the horizon only if, in addition, $\delta > (90° - \phi)$.

Figure 80 (*a*) shows the right-angled spherical triangle *ZPX*, *Z* being vertically above the observer's position *A*. *PA* is the direction of true north from *A*, and therefore the azimuth of *B* from *A* is the angle *PZB*.

The given sides are:

co-declination, $z = 90° - 61° 57′ 24″ = 28° 02′ 36″$
co-latitude, $x = 90° - 43° 58′ 12″ = 46° 01′ 48″$

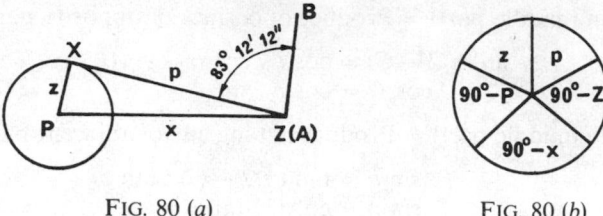

FIG. 80 (*a*)   FIG. 80 (*b*)

Taking z as the middle part, Napier's rule (1) gives:

$$\sin z = \cos (90° - Z) \cos (90° - x) = \sin Z . \sin x$$
$$\therefore \ \sin Z = \sin z/\sin x$$

Using 7-figure logs:

$$\log \sin z = \log \sin 28° \ 02' \ 36'' = 9.672\,226\,4$$
$$\log \sin x = \log \sin 46° \ 01' \ 48'' = \underline{9.857\,153\,5}$$

Subtracting, $\log \sin Z \ \ = 9.815\,072\,9$

$$\therefore \ Z = 40° \ 47' \ 11''$$

add $\underline{83° \ 12' \ 12''}$

Azimuth of B from $A = \underline{\underline{123° \ 59' \ 23''}}$

Other methods of determining azimuth

In Specimen Question 45 and 46 three methods of determining the azimuth angle of the celestial triangle have been outlined, namely:

(*a*) by circum-polar star at elongation;

(*b*) by extra-meridian observation on a star by measuring the altitude; and

(*c*) by extra-meridian observation on a star by measuring the hour angle.

The sun, of course, is much more easily identified than any star and can, too, be observed during daylight—i.e. in normal working hours—and for these reasons observations on the sun are often preferred by engineering surveyors. Because it is so much nearer than the stars, observations on the sun have to be corrected for parallax (for this correction, the *Star Almanac* states that "it suffices to add 0'.1 to all altitudes less than 70°") and have to be taken on both limbs (*see* Fig. 81).

The sun's declination (and other properties) are given in the first table of the *Star Almanac*. A celestial triangle can be drawn with vertices at the celestial pole, the sun and the zenith, and consequently two more methods of determining azimuth are:

(*d*) by extra-meridian observations on the sun by measuring the altitude; and

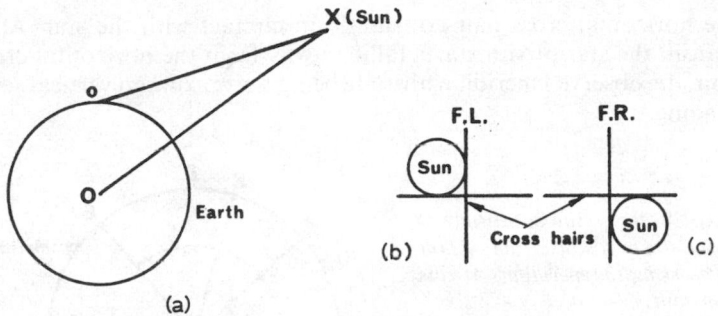

FIG. 81. *Correction for parallax.* (a) *Angle oXO is the correction for parallax. Adding this correction to the altitude measured from o yields the altitude of the sun reckoned from O, the centre of the earth, so that one set of tables may be used by observers anywhere on earth.* (b) *and* (c) *show the observations usually made on the sun (in rapid succession) for the best accuracy. The mean reading of horizontal and vertical scales can be taken to give the position of the sun's centre at the mean time of observation.*

(e) by extra-meridian observations on the sun by measuring the hour angle.

All five methods listed above depend on knowing the observer's latitude and having tables of declination available. If the latitude is not known it can either be determined separately or one of the remaining methods can be used:

(f) by using Pole Star tables (*see* Specimen Question 44).

(g) by equal altitudes of a star. A star has its maximum altitude at upper culmination, i.e. when it crosses the meridian. If, therefore, the horizontal scale of the theodolite is read each time the star reaches a certain altitude (given by the vertical scale reading), the mean horizontal reading will give the meridian, i.e. the direction of true north.

These methods and the methods of determining latitude are summarised in Table 12, in which the appropriate formulae are also given. It should be realised that other methods, not included, also exist.

METHODS OF DETERMINING LATITUDE

Figure 82 shows the most convenient sketch for latitude calculations. Evidently one method of determining latitude is:

Meridian altitude of a star or the sun
If the star (or the sun) is sighted through a theodolite telescope a few minutes before upper culmination, it will still be rising and can be followed by the telescope by turning the vertical tangent screw to keep

the horizontal cross hair constantly in contact with the star. After transit, the star or sun starts falling away from the horizontal cross hair, the observed meridian altitude being the maximum vertical scale reading.

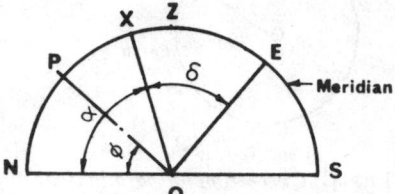

FIG. 82. *Determining latitude. X is the sun or a star at the instant of crossing the meridian. At this instant,* $\phi + 90° = \alpha + \delta$, *i.e. latitude,* $\phi = \alpha + \delta - 90°$.

Latitude can also be determined by sighting a suitable star at lower culmination.

Whether upper or lower culmination is to be used, it is not necessary to calculate the time of transit if the surveyor is prepared to spend, possibly, many hours observing one star waiting for it to cross the meridian. In practice, the time of transit is calculated roughly in order to reduce the fieldwork to a few minutes' duration.

Meridian altitudes of a zenith pair of stars

This is another method of determining latitude. A zenith pair is two stars, such as X_1 and X_2 (Fig. 83), which cross the meridian at nearly the same time and which, at transit, are roughly the same distance from Z (i.e. $\alpha_1 \simeq \alpha_2$). There are, in all, four different cases, since, by suitably redrawing Fig. 83, X_1 can lie on either NP or PZ, and X_2 can lie on ZE or ES. It is left as an exercise for the reader to draw the other sketches and to derive the expression for ϕ in each case, bearing in mind that δ is measured from the equatorial plane and is always less than $90°$, while α is measured from the plane of the horizon and is always less than $90°$. In all four cases the expression for ϕ is of the form:

$$\phi = \tfrac{1}{2}(\alpha_1 - \alpha_2) + \text{a function of } \delta_1 \text{ and } \delta_2$$

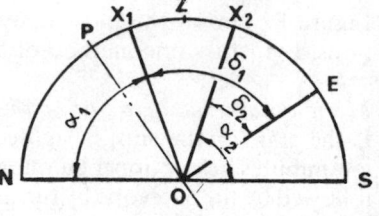

FIG. 83. *Meridian altitudes of a zenith pair of stars.* X_1 *and* X_2 *are a zenith pair of stars.*

$$90 + \phi = \alpha_1 + \delta_1$$
$$90 - \phi = \alpha_2 + \delta_2$$
$$\therefore 2\phi = (\alpha_1 - \alpha_2) + (\delta_1 - \delta_2)$$
$$\text{and } \phi = \tfrac{1}{2}(\alpha_1 - \alpha_2) + \tfrac{1}{2}(\delta_1 - \delta_2)$$

The declinations δ_1 and δ_2 can be obtained from the *Star Almanac*, while α_1 and α_2 are measured by theodolite. The method has the advantages that, since ϕ depends on the difference of α_1 and α_2, which are nearly equal, instrument errors will nearly cancel, as will refraction corrections (*see* below).

Suitable zenith pairs for different latitudes are given in the *Star Almanac* but attention has to be paid to the brightness of the stars before a choice is made.

Pole Star Tables: see Specimen Question 44. Latitude may also be determined by reference to these tables.

NOTES ON FIELD PROCEDURE

All the methods in this chapter have been described in the simplest possible manner. In practice, these simple procedures have to be elaborated in order to eliminate instrument errors, refraction errors and errors due to other causes.

Instrument errors

In a well-adjusted theodolite the main sources of instrument errors are:

(*a*) *Vertical scale index error*, which can be eliminated by taking face-left and face-right observations.

(*b*) *Movement of the altitude bubble*, for which allowance can be made by reading the objective and eyepiece ends (i.e. both ends) of the bubble tube at every pointing of the telescope.

Altitude

Altitudes are classed as:

(*a*) *observed* altitude = value read on the vertical scale;

(*b*) *apparent* altitude = observed altitude + instrument correction + semi-diameter correction (if necessary, e.g. for sun observations);

(*c*) *true* altitude = apparent altitude + refraction correction + parallax correction (if necessary).

Refraction correction, *r*

This is obtained from tables in the *Star Almanac* and is explained on p. x of the Introduction therein.

SPECIMEN QUESTION 47

Determine the refraction correction to an apparent altitude of $43° 57' 12''$ taken when the barometer read 691 mm of mercury and the temperature was $15°$ C.

TABLE 12. METHODS OF DETERMINING AZIMUTH AND LATITUDE

	Method	Figure	Available data — Known	Available data — From Star Almanac	Available data — Observed	Formula
AZIMUTH	1. Elongation of circumpolar star	(a)	$X = 90^\circ$, ϕ	δ	θ	$\sin Z = \cos \delta/\cos \phi$ (Napier's rule)
	2 and 4. Extra-meridian observation on star or sun (by altitude)	(a)	ϕ	δ	$\alpha,\ \theta$	$\tan (Z/2) = \sqrt{\dfrac{\sin\{s-(90^\circ-\phi)\}\ \sin\{s-(90^\circ-\alpha)\}}{\sin s\ \sin\{s-(90^\circ-\delta)\}}}$
	3 and 5. Extra-meridian observation on star or sun (by hour angle)	(a)	ϕ	δ	$P,\ \theta$ (by chronometer, and R.A.)	$\tan \tfrac{1}{2}(Z+X) = \dfrac{\cos \frac{1}{2}\{(90^\circ-\delta)-(90^\circ-\phi)\}}{\cos \frac{1}{2}\{(90^\circ-\delta)+(90^\circ-\phi)\}}\ \cot P/2$ $\tan \tfrac{1}{2}(Z-X) = \dfrac{\sin \frac{1}{2}\{(90^\circ-\delta)-(90^\circ-\phi)\}}{\sin \frac{1}{2}\{(90^\circ-\delta)+(90^\circ-\phi)\}}\ \cot P/2$
	6. Polar Star tables	(a) (c)		a_0, a_1, a_2 b_0, b_1, b_2	$\alpha,\ \theta$	$\phi = \alpha + a_0 + a_1 + a_2$ $Z = (b_0 + b_1 + b_2)\ \sec \phi$
	7. Equal altitudes of a star	(b)	$\alpha_1 = \alpha_2$	—	$\theta_1,\ \theta_2$	Azimuth of R.O. $= \tfrac{1}{2}(\theta_1 + \theta_2)$
LATITUDE	8. Meridian altitude of star or sun	(c)	ϕ	δ	α	Simple arithmetic
	9. Zenith pair of stars	(c) (modified)		δ_1, δ_2	α_1, α_2	Simple arithmetic
	10. Pole Star tables	see Method 6				

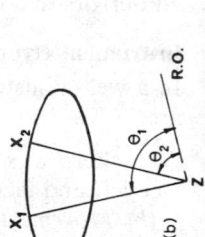

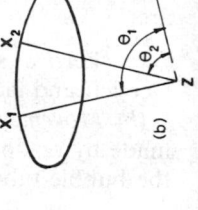

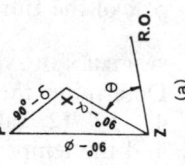

SOLUTION

From the appropriate tables of the *Star Almanac*, $r_0 = 60$ and $f = 0.89$

∴ Refraction correction, $r = r_0 \times 60 \times 0.89 = 53.4''$

i.e.   Refraction correction, $r = $ (say) 53 seconds of arc

Hence, assuming there is no parallax correction,

True altitude $= 43° 57' 12'' - 0° 00' 53'' = 43° 56' 19''$

Other terms used in field astronomy

(a) *Prime vertical:* the great circle through Z, at 90° to the meridian.

(b) *Ecliptic:* the path of the sun represented as a great circle on the celestial sphere, the plane of the ecliptic being inclined at 23° 27', approximately, to the plane of the celestial equator.

(c) *Obliquity:* the inclination of the Earth's axis (23° 27') to the normal to the plane of the Earth's path about the sun.

(d) *First point of Aries, γ:* the point at which the ecliptic crosses the celestial equator, the sun passing from south to north of the celestial equator. This event occurs at the spring (or vernal) *equinox.*

(e) *Right ascension* (R.A.): the angle between the meridian of a star and the meridian of $γ$ is the R.A. of the star. $γ$ is always considered to be ahead of the star; i.e. looking down on the celestial sphere from above the celestial north pole, so to speak. R.A. is always measured counter-clockwise from $γ$ to the star.

(f) *Precession of the equinoxes:* $γ$ moves roughly 50 seconds of arc along the celestial equator each year. This gradual movement is named the "precession of the equinoxes".

EXAMINATION QUESTIONS

1. A star of declination 47° 20' 17'' S is observed at upper transit at an altitude of 56° 48' 41'' N. What will be the observed altitude at lower transit and what is the latitude of the place of observation?

From this latitude, what will be the observed altitude of the top of the sun at transit if the declination of the sun is 22° 22' 25'' S and the sun's semi-diameter is 16' 18'''?

Explain why you could not use any of these observations alone to determine the azimuth of a survey line with any accuracy. What method would you use?

2. Using the formula (or any other one you may know) and information given below, compute the azimuth of the reference object from the observer's position.

Mean observed altitude (h) of the sun's lower limb in the E sky
$$= 36° 51' 43''.$$

Latitude (ϕ) = 51° 24' 00'' N.

Declination of sun (δ) = 5° 30' 00'' N and $p = 90° - \delta$.

Correction for refraction = 1' 14".
Correction for semi-diameter = 15' 24".
Correction for parallax = 07".
Horizontal angle measured clockwise from reference object to sun
$$= 86° 40' 30".$$

Formula:
$$\tan A/2 = \sqrt{\sec s \sin (s - h) \sin (s - \phi) \sec (s - p)}$$

where $s = \frac{1}{2}(h + \phi + p)$.

3. With the aid of diagrams and making use of the formulae

$$\sin \frac{A}{2} = \sqrt{\frac{\sin (s - b) \sin (s - c),}{\sin b \sin c}} \qquad 2s = a + b + c$$

show how the azimuth of a sun and a star can be obtained by the observation of altitude only.

What other information is necessary to solve the formula, and how would it be obtained? How precisely would one need to know the time, and for what purpose would it be required?

4. Define, with the aid of diagrams, the azimuth angle of the sun or a star.

Give three methods by which this angle can be determined in the northern hemisphere, stating what information is required and whether it is obtained by field observation or from an almanac.

5. Explain and indicate the significance of the following, in connection with astronomical observations: circumpolar star at elongation; star on the prime vertical; acceleration of sidereal on mean time; apparent time; correction tables for the determination of azimuth by Polaris.

6. Given the precise latitude and longitude of a station, how would you determine by astronomical methods the azimuth from that station to a chosen reference mark?

CHAPTER 11

AERIAL PHOTOGRAPHY

An aeroplane flying at about 4000 m and fitted with standard equipment for aerial surveying can produce a series of photographs each covering an area at ground level of over 3500 hectares. Some distortion will be inevitable but, nevertheless, the method clearly has advantages of speed and economy over traditional surveying methods of producing a plan of the same ground area.

Theorems on intersecting planes

As a preliminary to some of the bookwork for photogrammetry which follows, the reader needs to be familiar with certain concepts about intersecting planes. In the first place it has to be understood that a plane is infinite; when, for example, reference is made to "the plane of the photograph" this plane extends indefinitely in all directions and has no boundaries. A plane can be horizontal, vertical or inclined at any angle. Any three points in space define a plane; consequently three points are always co-planar (i.e. lie in the same plane) but this is not necessarily so for four or more points.

The theorems are as follows.

(a) Two parallel lines define a plane.

(b) Two planes always intersect in a straight line, provided that the planes are not parallel.

(c) Three non-parallel planes always intersect each other in a *single point* provided that one plane is not parallel to the line of intersection of the other two.

(d) Where two planes intersect, the line of maximum slope on one with respect to the other is at right-angles to the line of intersection of the two planes.

Vertical or oblique

Air photographs are described as vertical or oblique according as the optical axis of the camera is vertical or non-vertical.

It should be noted that, for a "vertical" photograph, the film is in fact horizontal at the moment of shooting.

Vertical air photography is by far the more common practice. As the camera is mounted in an aeroplane true verticality is seldom achieved but some obliquity, usually described in this connection as *tilt*, can be tolerated and the inaccuracies can usually be ignored if the tilt is less than about 3°.

Oblique photography is used only in very special circumstances, which are often of a military nature, because of its inherent disadvantage of dead ground, i.e. areas of the photograph where all the detail is obscured by a salient feature in front.

Single or stereoscopic pairs

Air photographs are normally taken one at a time, although *in toto*, they may form a series of photographs taken at regular intervals along the aeroplane's flight path. The title of this section, therefore, refers not to the way in which the photograph is taken but, indeed, to the way in which it is used after developing and printing.

Broadly speaking, single photographs can not be used for making detailed and accurate measurements and, therefore, lend themselves only to photointerpretation, a technique which, nevertheless, can yield very valuable information.

By using the photographs in pairs and viewing them stereoscopically, techniques are available which allow measurements to be made off them and plans to be drawn from them. The bulk of this chapter is devoted to these techniques and the underlying theory.

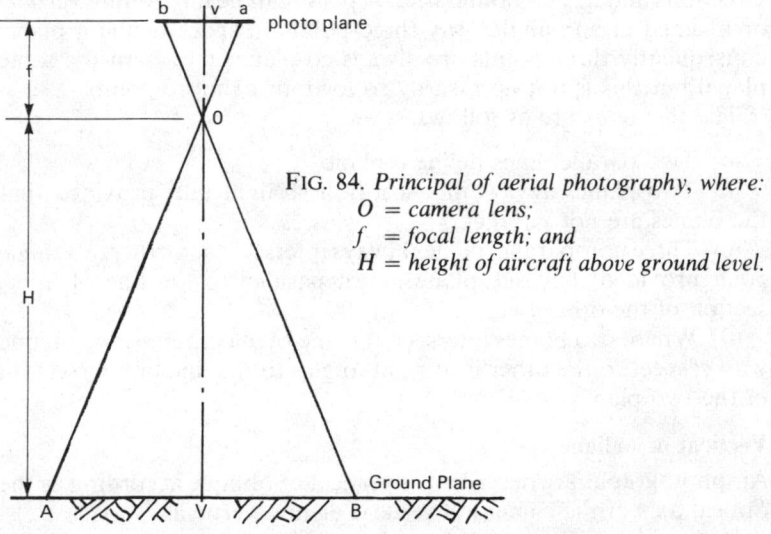

FIG. 84. *Principal of aerial photography, where:*
O = camera lens;
f = focal length; and
H = height of aircraft above ground level.

VERTICAL AIR PHOTOGRAPHS

Basic principle

In Fig. 84 the plane of the photograph (film or plate) is shown and so is the plane of the ground.

O is the camera lens;
f is its focal length;
H is its height above ground level;
AB represents the stretch of land being photographed;
ab represents the image of AB formed in the camera;
vV is a vertical line through O; and
v and V are known as *plumb points*.

In this simple example ab and AB are parallel and, therefore, $\triangle abO$ and $\triangle ABO$ are similar.

$$\therefore \frac{ab}{AB} = \frac{f}{H}$$

But $\dfrac{ab}{AB}$ is the *scale* of the photograph.

i.e. $\text{scale} = \dfrac{f}{H}$

A size of photograph commonly used is 230 mm square and it is also common for the camera to have a focal length of about 152 mm. Hence, taking $ab = 230$ mm, $f = 150$ mm and the flying height $H = 4000$ m, for example,

$$AB = 0.23 \times \frac{4000}{0.15} = 6130 \text{ m}$$

$$\therefore AB^2 = 6130^2 \text{ square metres}$$
$$= \underline{3760 \text{ hectares}}$$

which is an enormous area.

It should be noted that corresponding points such as B and b are called *homologous points*. Hence A and a are homologous and a is the *homologue* of A.

Distortions

The example used above to illustrate the basic principle is idealised. In practice, the ground is seldom perfectly level and it is not possible to ensure that the camera is perfectly vertical, both of which conditions lead to imperfections in the photograph. Un-level ground causes *distortion due to height* and a non-vertical camera causes *distortion due to tilt*.

Height distortion

This may be illustrated by considering the effect on the photograph caused by the vertical side of a tall building, BC, in Fig. 85.

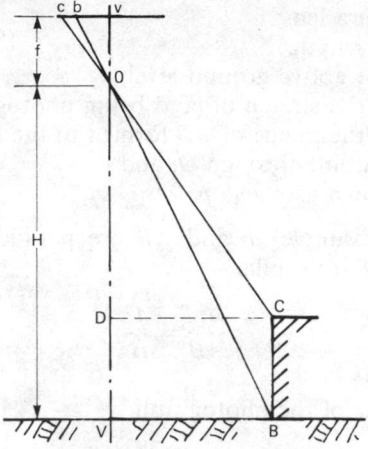

FIG. 85. *Height distortion.*

From a perfect overhead view *c* and *b* would coincide because *B* would be vertically below *C*. As Fig. 85 shows this is not the case, *b* and *c* being distinctly separate points on the negative.

The distance between *b* and *c* will depend both on the height, *BC*, and also on the distance of *B* from *V*. The length *bc* is, in a sense, a measure of the height distortion.

SPECIMEN QUESTION 48
The following information was recorded on an air photograph:

f = 300 mm, altitude = 600 m, scale 1:2000

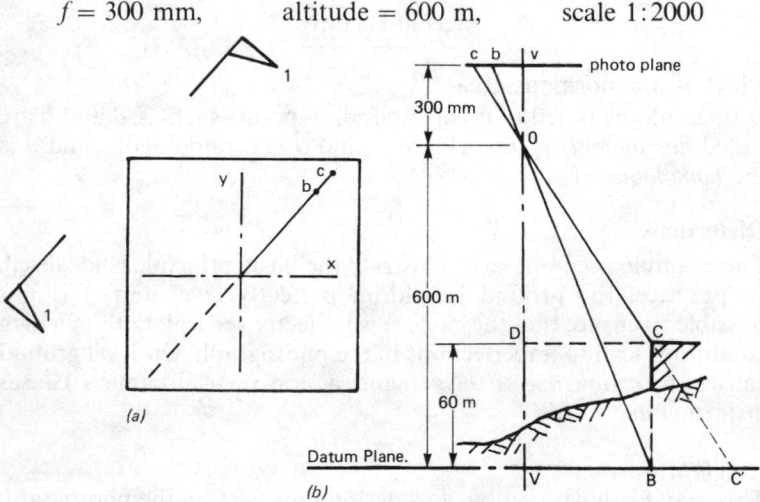

FIG. 86 (*a*) *The air photograph in Specimen Question 48.* (*b*) *View on* 1 − 1.

and these figures were used to plot the position of a building. The co-ordinates of the plotted position were $x = 165$ m, $y = 220$ m, the origin of the system of co-ordinates being taken at the plumb point. Unfortunately two mistakes were made in the plot in that (a) the co-ordinates scaled off the photograph were measured to the top of the building's image and (b) no allowance was made for the R.L. of the ground which, at the foot of the building, was 40 m above datum.

If the building was 20 m high calculate the true plan co-ordinates of its base and evaluate the error in its plotted position.

SOLUTION

The scale $\dfrac{1}{2000}$ applies only to points in the datum plane.

The plotted point for the building is, therefore, C' when it should have been B (c and b respectively on the photograph) and $BC' = $ error (see Fig. 86). Intuitively, it would seem the error lies along the radius from V (or v on the photograph) and this is proved on p. 178. The scale which should have been used, from the heights on Fig. 86, is

$$\frac{vO}{OD} = \frac{0.300}{600 - 60} = \frac{1}{1800}$$

Scale actually used $= \dfrac{1}{2000}$ and $\therefore$ correction factor $= \dfrac{9}{10}$

$$VC' = \sqrt{165^2 + 220^2} = 275 \text{ m}$$

$$\therefore VB = \frac{9}{10} \times 275 \qquad = 247.5 \text{ m}$$

$$\therefore \text{Error} = BC' \qquad = \underline{27.5 \text{ m}}$$

Correct co-ordinates are $\left(\dfrac{9}{10} \times 165, \dfrac{9}{10} \times 220\right) = \underline{\underline{(148.5, 198)}}$

It should be noted that the same answers will be obtained if the problem is solved by applying the properties of similar triangles direct and the reader is recommended to do this.

Tilt distortion

In Fig. 84 the photo plane is drawn parallel to the ground and may be said to be horizontal. If the camera were tilted the photo plane will not be horizontal and the image will be distorted, see Fig. 87 for example, in that when $AV = VB$, av and vb will not be equal and, therefore, the scale will be varying across the negative. This effect is called *tilt distortion* and is dealt with more fully in Specimen Question 49.

SPECIMEN QUESTION 49

(a) Define the following terms used in connection with aerial photography: Principal Points, Principal Plane, Plumb Points, Isocentres, Horizon Trace and Plate Parallel.

(b) Explain how, in practice, the position of the Principal Point can be located on an air photograph.

(c) Explain what is meant by the statements:

height distortions are radial from the plumb point;

tilt distortions are radial from the isocentre,

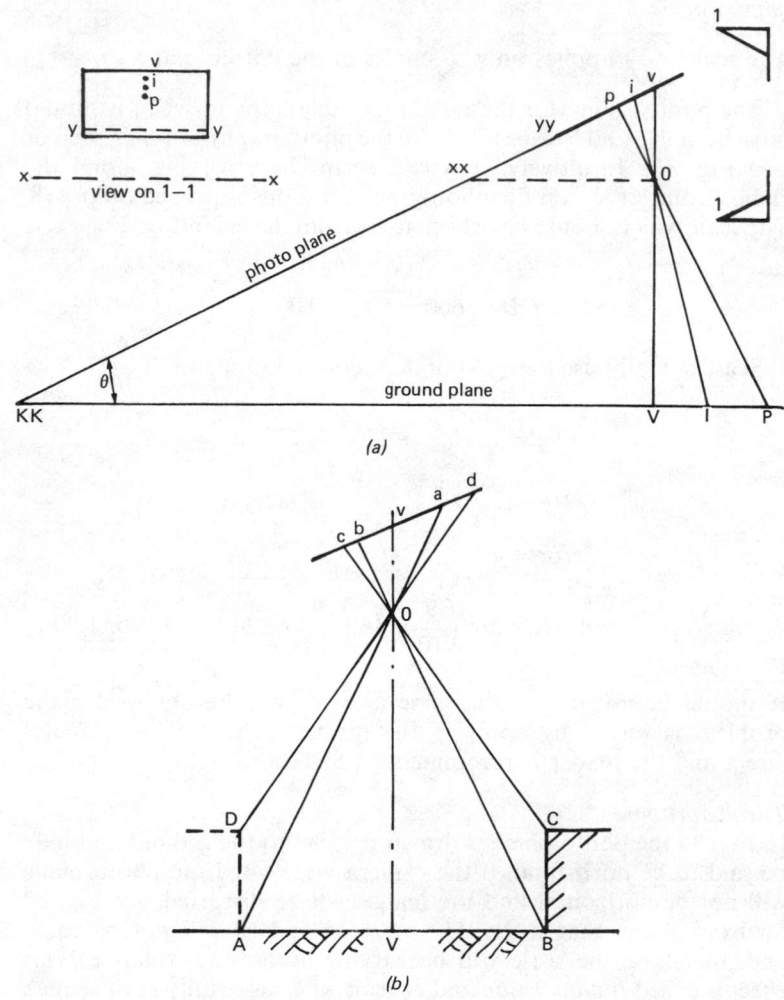

FIG. 87. *Tilt distortion.*

and state under what conditions distortions may be considered to radiate about the Principal Point.

(a) Figure 87 (a) shows in one view a section through the photograph and ground planes and projected from it a view of the negative from 1–1

$$\theta = \text{angle of tilt}$$
$$O = \text{camera lens}$$

p = principal point of the photograph and is at the centre of the photograph. pOP is therefore at 90° to the photo plane and P is called the principal point on the ground plane.

vV are the plumb points, see p. 173, and vOV is at 90° to the ground plane. From the geometry of the figure $V\hat{O}P = v\hat{O}p = \theta$

iOI is drawn to bisect $V\hat{O}P$ (and $v\hat{O}p$) and defines the two isocentres;

hence $$V\hat{O}I = I\hat{O}P = v\hat{O}i = i\hat{O}p = \frac{\theta}{2}$$

KK is the line along which the photo plane and ground plane intersect and, since Kp is a line of maximum slope, KK is at 90° to Kp and KP. The plane containing Kp and KP (i.e. the plane of the drawing) is called the Principal Plane and is usually described as the plane containing $OVIPvip$. A horizontal plane through O intercepts the photo plane in xx; xx is called the Horizon Trace and any line parallel to it, such as yy, is called a Plate Parallel.

(b) When aerial photographs are printed they bear fiducial marks by means of which the Principal Points can be located. These marks are often small vees, one let into the print on each side of the photograph, or something similar; lines joining opposite ones intersect at p.

(c) Height distortion and tilt distortion have already been described. In Fig. 87 (b) there would be no height distortion if BC were along OV; i.e. the height distortion is zero at the Plumb Point. For every other position of BC there will be height distortion and it is proved on p. 178 that it will always lie along the radius from V to BC, which is the same as the radius joining v to bc. Height distortion is said, therefore, to be radial about the Plumb Point.

In the case of Tilt it is shown on p. 179 that, along iOI the scale is $\frac{f}{H}$, the same as for a vertical photograph. On every other line through O the scale has a different value, owing to tilt distortion, and it is proved on p. 179 that it will always lie along a radius from the axis iOI, i.e. tilt distortions are radial from the isocentre.

When the camera is truly vertical, P, I and V coincide and p, i and v coincide. Provided that the angle of tilt is small, say $\theta < 3°$, P, I and V are very close. Therefore, for simple plotting, the assumption can be made that all distortions are radial about the Principal Point.

N.B.: the use of the radial line plotter, p. 188 is based on this assumption.

Proof that height distortion is radial about plumb point

Figure 88 shows an elevation, in which XY represents a height which is not in the principal plane, and also shows a plan view, in projection, to make this clear.

OV and XY by definition are vertical and therefore define a plane. Since YO and VO lie in this plane, so must y and v because YOy and VOv are rays of light passing through the centre of the camera lens.

$\therefore OVXYv$ and y are co-planar and x also must lie in this plane. But xy is the height distortion which therefore radiates from v.

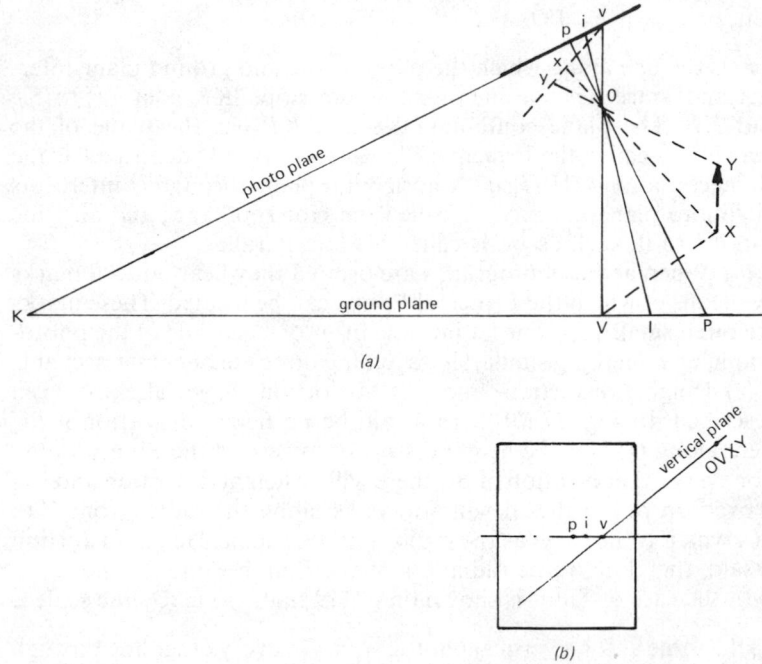

FIG. 88. *Height of tilt distortion.* (*a*) *elevation* (*b*) *plan.*

Proof that tilt distortion is radial about the isocentre

Figure 89 is a three-dimensional sketch in which X, x do not lie in the Principal Plane (i.e. $PIVK\ pivO$).

First, consider the scale factor along iOI:

$$pO = \text{focal length}, f, \quad \text{and} \quad OV = \text{height } H$$

$$\therefore Oi = \frac{f}{\cos \dfrac{\theta}{2}} \quad \text{and} \quad OI = \frac{H}{\cos \dfrac{\theta}{2}}$$

$$\therefore \frac{Oi}{OI} = \frac{f}{H}$$

which is the scale (*see* p. 173) when there is no tilt distortion. Nowhere else in this configuration does the scale have this value and, indeed, the factor decreases at points to the left of i (e.g. at p, scale $= \dfrac{f}{H} \cos \theta$) and increases to the right of i (e.g. at v, scale $= \dfrac{f}{H \cos \theta}$).

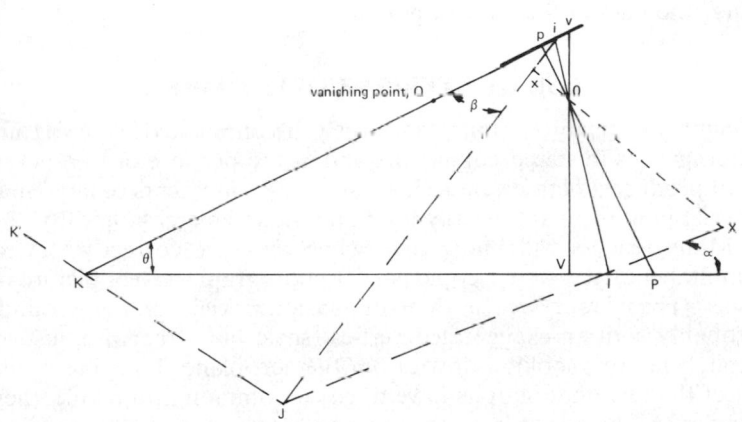

FIG. 89. *Diagram for the proof that tilt distortion is radial about the isocentre.*

IXO define a plane and since XOx and IOi are straight $OIXi$ and x are co-planar. There are, therefore, three intersecting planes: $OIXix$ the photo plane and the ground plane and if these intersect at J, J must lie on $K'K$ (*see* plane theorems, p. 171). Also $i\hat{K}J = 90°$ and $I\hat{J}K = 90°$.

From the geometry of the Principal Plane $\triangle\, iKI$ is isosceles with

$$K\hat{i}I = K\hat{I}i = 90° - \frac{\theta}{2}$$

$$\therefore KI = Ki$$

and it follows that KiJ and kIJ are congruent (2 sides and a right angle, KJ being a common side). Hence $\alpha = \beta$.

But the scale distortion for point X is $\dfrac{Ox}{OX}$ and since Ox and OX lie on the same radial plane through iI the tilt distortion is radial about iI.

Rectification

Where distortions due to tilt occur the negative can be rectified by, in effect, setting it up at its angle of tilt and rephotographing it on to a horizontal surface. Alternatively, it can be mounted horizontally and projected on to a suitably tilted plane. Special machines are available to facilitate rectification.

The process does not rectify height distortions, however. For very uneven ground corrections in this respect can be obtained only if ground control points have been established and accurately fixed by ordinary ground surveying methods. The position of the topographical detail shown on the photograph can then be modified so that it "fits" the plan of the control points.

USE OF STEREOSCOPIC PAIRS

Height and tilt distortions can largely be eliminated by using air photographs in stereoscopic pairs and, if the purpose of the survey is to produce a plan or, indeed, accurate measurements of any kind, overlapping pairs are usually used and viewed stereoscopically.

Many readers will know that, when they are correctly viewed simultaneously, a stereoscopic pair of photographs give a picture in relief. The viewer sees a three-dimensional view of the ground, probably with an exaggerated vertical scale but, otherwise, just as though he were looking down from the aeroplane. To achieve this effect the two photographs have to cover common ground (i.e. they overlap) both have to be to the same scale and, also, they have to be correctly oriented with respect to one another. In practice, distortions will be present, to some extent, if the cameras were not truly vertical.

If these conditions are fulfilled a feature such as BC (which is shown as an arrow only to indicate height and sharpness) in Fig. 90 (a) will appear on both negatives, the image being b_1c_1 on one and b_2c_2 on the other. There is, therefore, height distortions on both. Figure 90 (b) represents what is seen when the photographs are viewed stereoscopically; namely the image bc, in relief, at the point of intersection of the two rays. Notice that, to be able to do this, both Principal Points have to be marked on each print, it being fairly easy to transfer a principal point from one photograph to the corresponding position on the other.

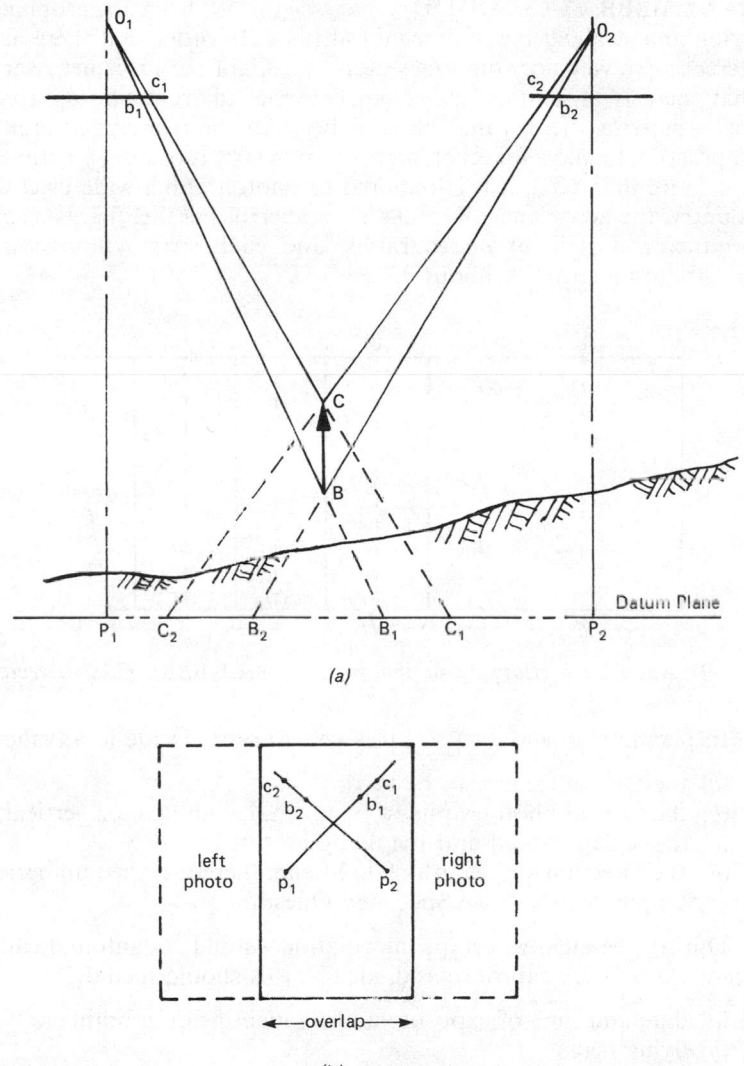

FIG. 90. *A pair of air photographs, (a) being taken and (b) positioned for stereoscopic viewing.*

How overlapping pairs are obtained

In practice, it is usual to take a series of overlapping photographs AAAA, BBBB, CCCC, DDDD ... etc., *see* Fig. 91, from an aeroplane flying in a straight line at constant altitude. In order that there are stereoscopic views continuously along the flight path it is necessary that there is at least 50% overlap between adjacent photographs for, otherwise, there would be gaps between the overlapped areas. In practice, to allow for error, an overlap of 60% is usual. Sometimes it is more than 60%. If it is required to photograph a wide tract of country, the aeroplane will make a number of parallel flights, each producing a strip of photographs, and each strip will overlap neighbouring strips by about 25%.

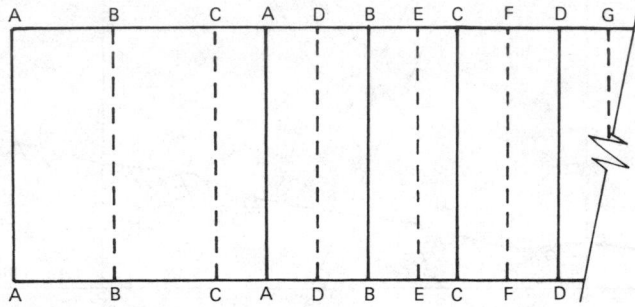

FIG. 91. *A series of overlapping air photographs AAAA, BBBB, CCCC.... etc.*

In planning an aerial survey it is necessary to decide in advance:

(*a*) the type of camera to be used;
(*b*) the type of photographs to be taken, i.e. oblique or vertical;
(*c*) the scale required and the flying height;
(*d*) the direction and length of flight and, therefore, the number of photographs required, *see* Specimen Question 50.

During the survey certain information should be automatically recorded on each exposure and, ideally, this should include:

(*a*) date and time of exposure and an identification number;
(*b*) flying height;
(*c*) a spirit level reading;
(*d*) a scale factor;
(*e*) fiducial marks for locating the Principal Point.

An instrument box is associated with the camera to provide some or all of this information. In addition the sighting device comprises a telescopic view-finder through which can be seen two reticules, one showing the ground coverage and the other facilitating overlap control by means of reference lines which are synchronised to move with the passing image of the ground.

SPECIMEN QUESTION 50

(a) Explain why the overlap of stereoscopic pairs of photographs is normally 60%.

(b) Calculate the number of exposures required in one strip of 11 km in length if the scale is to be 1 : 4800 and the photographs 230 mm square. If the area to be surveyed were 5 km wide, calculate the number of parallel flights to be made by the aeroplane.

(c) Distinguish between uncontrolled and controlled mosaics and explain under what circumstances the overlap may be increased beyond the normal 60%.

SOLUTION

(a) *See* above, p. 182.

(b) Referring back to Fig. 91

Overlap of one stereoscopic pair = 60%
Of this 20% is common to the overlap of the next pair.
Hence contribution of one pair = $40\% \times 230 = 92$ mm
i.e. each photograph contributes 92 mm;
On the ground, scale equivalent = $4800 \times 0.092 \doteq 442$ m

$$\therefore \text{No. of exposures} = \frac{11}{0.442} + 1 = \underline{\underline{26}}$$

NOTE

1 is added to the calculated value because, of course, two photographs are needed to give one overlap, three photographs to give two overlaps ... and so on.

Taking the overlap between strips to be 25%, the contribution of one strip

$$= 0.75 \times 0.23 \times 4800 = 828 \text{ m}$$

$$\text{No. of strips} = \frac{5000}{828} = 6.04, \text{ say } \underline{7}$$

(c) When an aerial survey is performed over an area such as that in (b) above, an alternative way of using the photographs is to mount them on a plane backing to form a continuous photographic picture of the whole area. Photographs mounted in this way form what is called a mosaic. If every photograph is rectified and brought to a common scale, the mosaic is said to be controlled. If, however, the prints are mounted unmodified, the mosaic is described as un-controlled and will exhibit tilt and height distortions and a non-uniform scale—and is inferior, therefore, to controlled mosaics.

In controlled mosaics, height distortions are sometimes minimised by trimming each rectified photograph to leave only the central parts which are then joined to form the mosaic.

Large scale mosaics are sometimes obtained by enlarging a single air photograph. When this is the intention the flight planning may be modified to increase the overlap to as much as 90%.

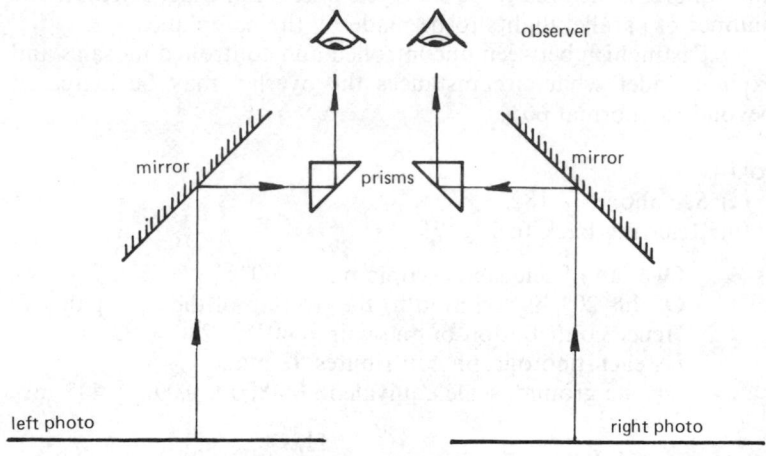

FIG. 92. *Principle of the mirror stereoscope.*

STEREOSCOPIC VIEWING OF PHOTOGRAPHS

To get the 3-D effect an overlapping stereoscopic pair of photographs have to be viewed simultaneously, one with each eye. Although it is possible, with practice, to do this without any optical aid, it is much more easily accomplished with the aid of a stereoscope.

With or without optical aid successful viewing can be accomplished only if the pair of photographs are carefully and correctly position. Normally the Principal Point on each photograph is located (from the fiducial marks) and marked. To each photograph is added the Principal Point from the other member of the pair by, for example, pricking through. If these are labelled p_1 and p_2, the line joining p_1 and p_2 is called the air base and this must be parallel to the line joining the viewer's eyes. The photographs are placed side by side so that the distance separating the two points marked p_1, say (or any other pair of conjugate points which appear one on each photograph) is about the same as the distance between the eyes. It is also important that the photographs are the right way round, namely, so that the "left" photograph is viewed by the left eye.

A simple pocket stereoscope consists of two lenses, one for each eye, which are mounted on legs in a plane parallel to the plane on which the photographs are placed when being viewed. The lenses magnify the images slightly and enable the viewer to "read" the photographic detail more easily than he can without them. The double lenses also facilitate obtaining the three dimensional view.

A much more useful, but still relatively inexpensive, instrument is the mirror stereoscope, Fig. 92, which works on the same principle as a pocket stereoscope but which, by virtue of mirrors, allows the photographs to be separated by clear space (i.e. no overlap) when viewed. Two degrees of magnification are normally possible, one being obtained by inserting between the eye and the prisms two simple lenses, one for each eye, and the other by removing the simple lenses and inserting a pair of binoculars instead. These are supplied with the stereoscope and are purpose made to fit it. The mirror stereoscope can be used by itself for photo interpretation or in conjunction with the parallax bar, or stereometer, a device for carrying out simple height measurements. The mirror stereoscope forms the optical system used in the radial line plotter. Both of these instruments are described below.

METHODS OF MAKING MEASUREMENTS FROM STEREOSCOPIC PAIRS

The two operations described in this section are heighting and plotting.

Heighting is the term used to describe the process for determining, from a pair of overlapping photographs, the height of one point with respect to another: the height of a building or tree for example. In its simplest form it is carried out by means of a parallax bar.

Plotting is self-explanatory. Suffice it to say that it is possible to plot detail on to a plan or map direct from a stereoscopic pair by means of the radial line plotter or more elaborate plotting machines which are beyond the scope of this book.

Heighting
In Fig. 93 O_1, O_2 are the camera lenses. B = air base

XY represents a feature of height h above datum

For a stereoscopic pair of photographs, the parallax of a point such as x is defined as $p_2x_2 + p_1x_1$.

To make this expression easier to comprehend, the conjugate of x_2

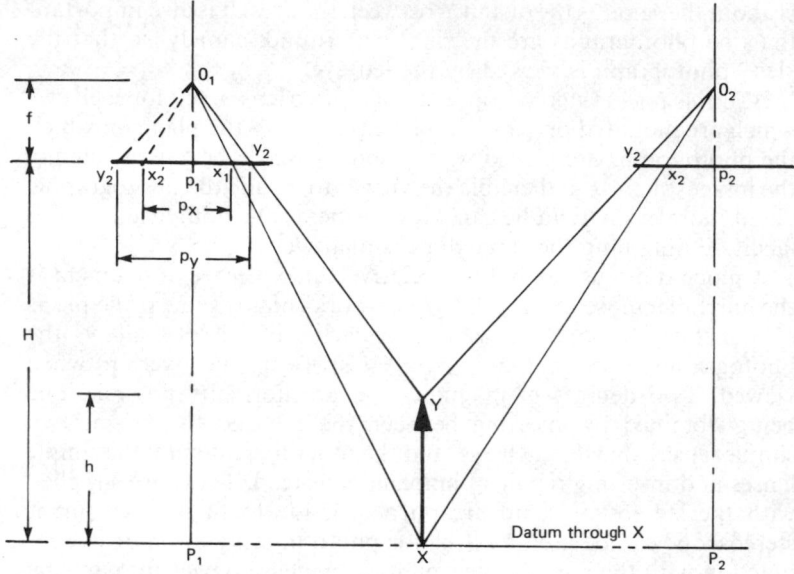

FIG. 93. *Diagram for "heighting".*

is marked on the left-hand positive and is labelled x_2' on the figure.
i.e. $p_2 x_2 = p_1 x_2'$. Similarly y_2' is the conjugate of y_2, so that $p_2 y_2 = p_1 y_2'$.

Let parallax of x be p_x

then $\qquad p_x = x_2' p_1 + p_1 x_1$ and, similarly,

parallax of y, $\quad p_y = y_2' p_1 + p_1 y_1$

$$\triangle O_1 O_2 X \text{ and } \triangle O_1 x_1 x_2' \text{ are similar and } \therefore \frac{p_x}{f} = \frac{B}{H}$$

$$\triangle O_1 O_2 Y \text{ and } \triangle O_1 y_1 y_2' \quad \text{,,} \quad \text{,,} \quad \text{,,} \quad \text{,,} \quad \frac{p_y}{f} = \frac{B}{H-h}$$

$$\therefore p_x = \frac{Bf}{H}$$

$$p_y = \frac{Bf}{H-h}$$

and most problems can be solved from these two equations used in
conjunction with the knowledge that the difference in parallax of P
and X can be measured in millimetres with a parallax bar.

i.e. $\qquad\qquad p_y - p_x = \text{parallax bar reading.}$

Also, from the equations for p_x and p_y

$$p_y - p_x = Bf\left(\frac{1}{H-h} - \frac{1}{H}\right) = Bf\left\{\frac{h}{H(H-h)}\right\}$$

$$\therefore \ h = (p_y - p_x)\frac{H(H-h)}{Bf} \simeq (p_y - p_x)\frac{H^2}{Bf},$$

$$\text{if } h \ll H$$

NOTE

In Fig. 93, to simplify the algebra, H is measured from the lower of the two points under consideration, X. In previous figures it has been measured from the map datum and these two definitions of H do not necessarily have the same value; they differ by the height of X above datum. In most cases the percentage difference between the two values will be small. Nevertheless the reader should understand that, by confusing these two definitions of H, an error might be introduced into the calculation.

The equations for p_x and p_y do not give the Reduced Level of a point. It must be emphasised that they can be used only for calculating differences in level. A Reduced Level can be determined only by referring height calculations to a nearby point of known Reduced Level, e.g. a benchmark.

The accuracy of the above equations depend on there being no tilt. As this is seldom the case, the calculations are unlikely to give a high degree of accuracy.

From the last approximate equation

$$\frac{p_y - p_x}{h} \simeq \frac{Bf}{H^2}$$

i.e.

$$p_y - p_x h \simeq \frac{f}{H} \times \frac{B}{H}$$

$$\simeq \text{horizontal scale} \times B/H$$

But $\dfrac{p_y - p_x}{h}$ is the *vertical* scale of the 3-D image seen through the stereoscope

$$\therefore \ \frac{vertical \ scale}{horizontal \ scale} \simeq \frac{B}{H}$$

Hence, as B and H are usually not equal, the vertical scale usually differs from the horizontal scale.

Parallax bar

A parallax bar comprises two cursors mounted near the ends of what is essentially a micrometer screw gauge. When a stereoscopic pair is viewed through the mirror stereoscope with which it is supplied, it is positioned so that a cursor lies over each of the pair of images of the feature under scrutiny. Each cursor carries a fine mark which might be either a cross, a dot or a small circle. By putting the parallax bar parallel to the line joining the Principal Points and turning the micrometer screw, the two marks can be made to fuse stereoscopically. They then appear to float over the picture and can be raised or lowered by turning the micrometer screw one way or the other. To measure the height of a building, say, the floating mark is first adjusted to be level with the foot of the building and the micrometer reading noted.

It is then raised to coincide with the top of the building and the new reading is noted. The difference between the two readings gives the difference in parallax measured to the nearest 0.01 mm. Experience and practice are needed to be able to do this competently.

The parallax bar has a coarse length adjustment so that it can be adapted to photographs of different sizes.

Plotting and the radial line plotter
The purposes of aerial surveying can be:

(*a*) to provide a pictorial view of the ground either "flat" or stereoscopically;

(*b*) to provide a computerised model of the ground which, when asked, yields, for example, co-ordinates giving cross-sections anywhere along the route of a survey—and this can be of considerable value to the Civil Engineer concerned with large-scale earth moving;

(*c*) for map production or up-dating.

Plotting is an essential part of the latter function. It has already been explained on p. 185 how points can be plotted from the intersection of rays radiating from the Principal Points on two photographs viewed stereoscopically. Moreover, plotting by this method reduces the grosser inaccuracies due to height and tilt distortion.

A simple machine designed to facilitate plotting is the radial line plotter which comprises a mirror stereoscope, two tables on which the photographs are placed for viewing and a pencil for marking the map. The Principal Point of each photograph is located on its table by means of a pin and this also provides the fulcrum about which a perspex radius arm, or cursor, can rotate. There are two cursors, one to each table, and it is their movements which, through a system of linkages, guide the mapping pencil. Most comprehensive surveying books show a photograph of a radial line plotter.

To use the machine:

(a) mark the Principal Points p_1 and p_2 say, at the centre of each photograph;

(b) mark the conjugate points p_1' and p_2' on each photograph and draw on the base lines $p_1 p_2'$ and $p_1' p_2$;

(c) carefully mount the photographs on their respective tables making sure that they are flat;

(d) insert the pins through the central principal points and attach the cursors;

(e) if the points to be plotted are called $a, b, c \ldots$ etc., carefully place one cursor over a_1 and the other over its corresponding point on the other photograph, a_2. The pencil then automatically marks the point a on the map and the process is repeated for $b, c \ldots$ etc.

The method works for all points which are not near the base line where the intersections are very oblique and, therefore, ill-defined. To overcome this difficulty, provision is often made for the radial arms to be used about false centres from where the plan can be completed.

With strips of photographs used in stereoscopic pairs, inaccuracies can accumulate as work progresses along the strip. To prevent this from happening it is necessary to establish a number of minor control points, well away from the Principal Points, on a continuous piece of tracing paper used as an overlay, so that any subsequent errors are localised. In short, by working "from the whole to the part". A more refined way of achieving the same end is to use slotted templates which is described more fully in, for example, *Land Surveying* by Ramsay J. P. Wilson.

In addition to the minor control points taken from the photographs, at least two good ground control points should be established with one near each end of the strip. The distance between them is accurately measured by ordinary ground surveying methods and the horizontal scale established.

More elaborate plotting machines

These are beyond the scope of this book. Suffice it to say that advanced use of aerial surveying techniques is a highly specialised industry calling for considerable experience and expertise in both the flying and the plotting.

PHOTO INTERPRETATION

Photo interpretation is the name given to the examination of air photographs in, broadly speaking, a non-quantitative way. It can refer to oblique photographs or vertical photographs, either singly or in stereoscopic pairs, and calls for a good deal of practice and expertise. It is a technique which can be of value to anyone concerned with the use of large stretches of land or sea as the following list indicates.

(a) *Admiralty* might use air photographs as an aid in charting navigation channels or for studying coastal waters.

(b) *Advertisers* may use obliques in their holiday brochures.

(c) *Archaeologists*, who are often concerned with microscopic examination of their finds, can benefit from the macroscopic examination of their sites by aerial survey for it is well known that large archaeological features can sometimes be seen as shadowy outlines on air photographs when nothing is apparent at ground level.

(d) *Civil engineers* might use air strips in their preliminary investigation of alternative routes of a proposed, new, major road or they may use air photographs in the preliminary considerations of coastal defence works or in choosing a new reservoir site.

(e) *Forestry, horticulture and agriculture*. Air photographs might be used to assess quantities of timber available in a plantation or, perhaps, the extent of crop disease.

(f) *Geographers and geologists* can learn from aerial photographs how land is used, possibly the general nature of the soil and, frequently, something about the geological history. Glacial moraines or ox-bow lakes are obvious examples.

(g) *Map makers*. Sub-continents can be mapped by aerial surveying in probably a quarter of the time that would be taken by more traditional surveying methods.

(h) *Military* can photograph enemy defences by oblique photographs without trespassing over enemy territory. More daring sorties over enemy territory can yield very informative vertical air photographs.

(i) *Traffic control* can be helped by air photographs, taken at appropriate times, to identify the obstructions which are causing traffic jams.

Correct interpretation of air photographs is not unlike detective work. The clues have to be observed and then pieced together to reveal the whole truth—this approach can only be touched on here in the limited space available. Whole books are devoted to the subject of photo interpretation and make a very interesting study for anyone equipped with a pocket stereoscope, a magnifying glass or even just the ability to view photographic pairs stereoscopically with the naked eye.

The features to be observed first are size, shape, shadow and pattern. Some buildings can be identified as industrial, rather than residential, by their size and shape or even by size alone if there are other clues available too. Trees can sometimes be identified or classed by their shadows and, although the wires in an overhead power line may not themselves be visible, their shadows may show clearly on the ground. The military might be able to identify camouflaged aerodromes and buildings by the shadows they cast or by the distinctive patterns of their runways. Major rail termini can be identified by the pattern of the lines or the over-all diverging shape of the lines fanning out as they do. Character can also give a clue. Snow, for example, has an unmistakable character; grasslands and other types of vegetation can, with practice, be identified; sports grounds might be recognised by the attitudes of the participants; industries by the type of plant and subordinate structures to be seen. When viewing stereoscopic pairs moving vehicles are not seen in relief but as two distinct "flat" images and the distance separating the two is an indication of speed. Railway tunnels can sometimes be traced by the line of chimneys built in the hill over the tunnel to let smoke escape in a bygone age.

The variety of clues is endless and can be appreciated only with the experience and enjoyment of viewing air photographs. An excellent book for a first introduction is *Aerial Photo Interpretation* by Barry Sully.

SPECIMEN QUESTION 51

(*a*) Explain how the parallax bar is used with a stereoscopic pair of air photographs for *heighting* and derive formulae from which the difference in level of two adjacent points can be calculated.

(*b*) The interpretation of air photographs is a skill which now has wide applications; briefly explain applications which might be of significance to the Surveyor.

SOLUTION

(*a*) Bookwork, *see* p. 185.

(*b*) Wherever extensive developments are envisaged aerial photographs, correctly interpreted, can convey environmental factors which may not be apparent from a map. For example, they might indicate whether a residential area is semi-rural or completely urban even to the extent of overcrowding; and would certainly reveal undulations of the land which may not show on a map. They would reveal much about the nature and extent of waste land or parkland, the suitability of existing or proposed industrial sites, or the availability of transport services. The adequacy of road systems might be revealed by air photographs taken in the rush hour.

They can be equally revealing about rural situations. They might show low-lying, badly-drained land or land which is too heavily wooded. The nature of farmland, e.g. arable, mixed, livestock or fruit farming would show. A proposal to convert on a grand scale, agricultural land to some other purpose, e.g. the extraction of minerals, might be examined by means of aerial photography to assess the feasibility of the proposal, the implications on local amenities and, to some extent, the geology of the area.

CHOICE BETWEEN AERIAL SURVEYS AND MAPS

Not all the advantages lie with aerial surveys and this chapter concludes, therefore, with a few observations on the respective merits of aerial surveys and maps.

Generally, aerial surveys are economically viable only for areas which measure many hectares in size. With apologies for the pun, the "overhead costs" are large for aerial surveys and the cost per unit area surveyed decreases as the over-all size of the survey increases. With ground surveys the reverse is the case and, therefore, beyond a certain size detailed ground surveys will become more costly than air surveys. This applies particularly in undeveloped countries for which neither maps nor air photographs exist.

In developed countries, such as Great Britain, maps do exist even though they are inevitably a little out of date. The choice here will depend on whether the advantages of an air survey are likely to lead to an economic gain which more than compensates for the cost of the air survey.

It should be understood that maps do yield information which either is not, or may not be, available from air photographs. For example, boundaries are marked on maps and certain buildings such as inns are labelled. Most maps label the major roads and some maps, i.e. street maps, label all the roads. Sometimes roads, railways or other features may be hidden from view in an air photograph, by a dense forest for instance, but are clearly marked on the map. Lastly, maps have the grid system for easy reference.

EXAMINATION QUESTIONS

1. The use of photography from the air for surveying purposes is now established. Discuss the uses to which such photographs may be applied. Explain how such aerial surveys or photographs are tied in to existing plans or information.

2. Explain clearly how heights may be measured from a pair of overlapping photographs using a stereoscope and parallax bar. Derive the formula necessary to convert the parallax bar readings into heights. What are the sources of error in measuring heights by this method?

3. Describe the distortions which exist on single aerial photographs and explain how these are overcome when plotting from an overlapping pair of photographs.

4. (a) Describe a camera sight as used in vertical air photography. Why is it necessary and how is it used?

(b) A strip of vertical photographs is to be taken along a straight line 20 km long in fairly flat country. The negative is 20 cm square and the photographs are required to a scale of 1/12000 with 60 per cent overlap. If the focal length of the camera is 200 mm, determine:

(i) the minimum number of photographs for the strip;
(ii) the height above ground at which the aircraft should fly;
(iii) the aircraft speed, relative to the ground, in order that the interval between exposures shall be 10 seconds.

5. (a) In connection with height determination from stereo pairs of aerial photographs explain why a parallax bar, or similar apparatus, does not normally give the absolute parallax of points on the photographs.

(b) Despite paragraph (a) above, the readings obtained from a parallax bar are useful in height determination; why?

(c) A pair of vertical photographs were taken with a camera of focal length 150 mm from a height of 3000 m above a datum plane. The parallax reading for a point M at datum level was 15.50 mm and for another point N, 17.65 mm. If the Principal Points are at datum level and their distance apart on one photograph is 53.50 mm, determine the difference in level between M and N.

6. (a) What is meant by (i) height displacement, (ii) tilt displacement in an aerial photograph?

(b) In what directions do these displacements lie?

(c) Explain the purpose and process of "rectifying" an aerial photograph.

7. The following information is printed on an aerial photograph:

$$\text{scale} \quad 1/10\,000, \text{ zero tilt, } f = 150 \text{ mm}$$

Taking the Principal Point as origin the co-ordinates of two points of detail H and K were measured as follows, x and y being the normal co-ordinate axes.

	x mm	y mm
H:	+50.0	+60.5
	−15.2	−40.0

The ground distance between was calculated on the assumption that the scale was as given on the photograph. It was not appreciated that the given scale applied only at a Reduced Level of 100 m whereas the R.L. of H was 210 m and the R.L. of K, 70 m.

Assuming the information on the photograph to be correct, determine the error in the calculated horizontal ground distance between H and K.

CHAPTER 12

THEORY OF ERRORS

In base line measurement, traverse tables and elsewhere, the need for applying corrections to compensate for errors has been demonstrated. In traverse tables relatively simple rules were given for distributing closing errors. In baseline measurement the errors are calculable and corrections are applied for slope, tension, temperature, etc. Nevertheless, there would be, in addition, small accidental errors because, in all surveying measurements, this is the case. If the survey is self-checking the closing errors will be manifested and some way of correcting them has to be found. At the very least these errors have to be distributed throughout all the observations and then eliminated so that no closing error remains. Otherwise, any subsequent calculations cannot be made consistently. For example, if the observed angles of a plane triangle do not add up to 180°–00′ 00″ trigonometrical calculations based on one angle will not be in accord with those based on another angle.

SPECIMEN QUESTION 52
A level run is carried out along the sides of a triangle ABC with the following results: $AC = +132.877$, $CD = -81.400$, $DA = -51.237$, + denoting a Rise and − a Fall. Distribute the closing error assuming all readings are equally weighted.

SOLUTION
Because all the readings are equally weighted, it is assumed that each reading is equally at fault.

Since the level run starts and finishes at A the sum of the Rises should equal the sum of the Falls. Hence:

closing error $= 132.877 - 81.400 - 51.237 = 0.240$ m

Each reading has, therefore, to be corrected by an amount $= \dfrac{0.240}{3}$ $= 0.080$, the sign of the correction being ascribed intuitively.

Corrected values are, therefore

AC:	$+132.877 - 0.080 =$	$+132.797$
CD:	$-(81.400 + 0.080) =$	-81.480
DA:	$-(51.237 + 0.080) =$	-51.317
Check:	Closing error $=$	0

In the above example the result would be different if there was good reason for assuming one of the readings was more accurate than the other. Suppose the measurement along CD had been performed twice and the recorded value of -81.400 were the mean of the two. Then it would be considered to be twice as reliable as the other two readings and, therefore, it would seem reasonable to assume that, if the error in $AC = e$ and the error in $DA = e$, the error in CD would be $\frac{1}{2}e$. Corrected readings could be set out as follows, therefore

Line	Rise	Fall
AC	$132.877 - e$	
CD		$81.400 + \frac{1}{2}e$
DA		$51.237 + e$

Closing error $= 132.877 - e - (81.400 + \dfrac{e}{2} + 51.237 + e) = 0$

i.e. $\qquad \dfrac{5}{2}e = 0.240$

$$e = \frac{2}{5} \times 0.240 = 0.096 \quad \text{and}$$

Corrected readings would be

AC:	$+132.877 - 0.096 =$	132.781
CD:	$-(81.400 + 0.048) =$	81.448
DA:	$-(51.237 + 0.096) =$	51.333
Check:	Closing error $=$	0

If one or more observations are likely to be more reliable than the others, they are said to be weighted and, in the above calculation, the weight of $AC = 1$, $DA = 1$, but $CD = 2$. Calculations with weighted values are not necessarily set out in this way but this simple example illustrates the concept of weighting.

Other factors may further complicate the process of distributing the errors.

For example, the round of levels $ACDA$ above may be part only of a more complicated levelling exercise $ABCDA$, Fig. 94. If the difference in level along AB and BC have also been measured, there are three possible circuits, $ABCA$, $ACDA$, and $ABCDA$, round each of which the closing error has to be zero. This case calls for more advanced techniques described later in the chapter, which are all based on the theory of Normal Distribution of errors.

NORMAL DISTRIBUTION

Wherever physical or mathematical quantities are known to vary in a random fashion within narrow limits it is possible to calculate a

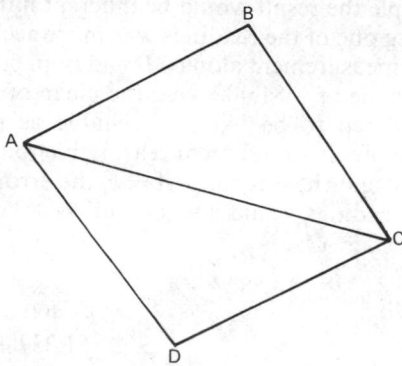

FIG. 94. *Round of levels.*

most probable value by applying a knowledge of statistics. It is not possible in every case to formulate a mathematical theory to simulate exactly the random variation but experience shows that the so-called normal distribution approximates very closely to that obtained in a wide range of experimental results and, in fact, the random errors occurring in surveying observations are usually assumed to vary in this way.

A comprehensive treatment of normal distribution theory is very mathematical and below, therefore, only the main features are described.

The normal distribution curve

The equation in its most general form is

$$y = \frac{1}{\sigma\sqrt{2\pi}} \; e^{-\frac{1}{2}\left(\frac{u-\mu}{\sigma}\right)^2}$$

the curve being shown in Fig. 95 (*a*). If a measurement of magnitude u were to be made a large number of times and if y, the number of times that each particular value of u was obtained, is also recorded, then a plot of u v$_s$ y would lie on this curve. (In practice, a histogram is the nearest we can get to the theoretical curve.) Moreover the mean value of all the u observations would be μ and the line $u = \mu$ would be the axis of symmetry. σ is not so easily portrayed but, in fact is a measure of the spread of u—values about the mean, μ. σ and μ are parameters which control the shape, size and position of the curve on the graph. Figure 95 (*b*) shows three curves, each of which corresponds to a different pair of values of σ and μ.

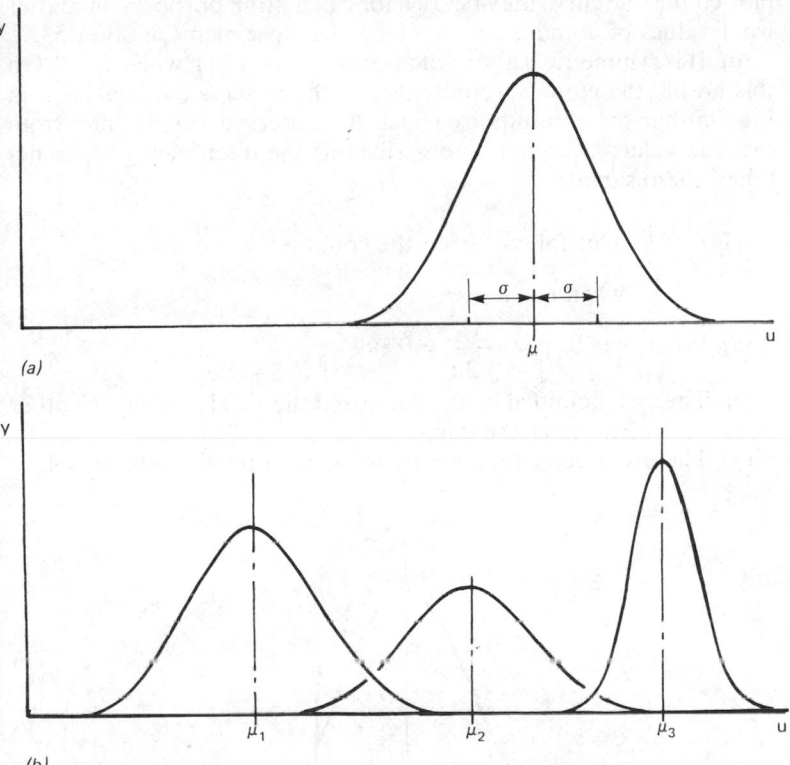

FIG. 95. (a) *The normal distribution curve.* (b) *Normal distribution curves corresponding to different pairs of values of σ and μ.*

Standardised normal distribution

By making a substitution in the above equation of $x = \dfrac{u - \mu}{\sigma}$ the equation becomes

$$y = \frac{1}{\sqrt{2\pi}}e^{-\frac{x^2}{2}}$$

which is a tidier expression, is easier to remember but, mathematically, is not really any simpler than the general equation.

In this form, the equation is said to be standardised and the curve it gives is plotted in Fig. 96. This is a much more convenient form to use in calculations because of its properties which are given below.

(a) The curve is independent of μ and σ; in other words once it is plotted the one curve may be used for calculation purposes no matter what values of μ and σ are involved. *See* Specimen Question 53.

(*b*) It is symmetrical about the *y*-axis. Hence $u = \mu$ when $x = 0$. On this graph, therefore, x represents not the observed values (as u in Fig. 95) but the amounts by which the observed values differ from the true value; i.e. x can be described as the discrepancy or, sometimes, the residual.

(*c*) $\sigma = 1$. This follows from the equation $x = \dfrac{u - \mu}{\sigma}$

when $u - \mu = \sigma$.

(*d*) When $x = 0$, $y = \dfrac{1}{\sqrt{2\pi}} = 0.399$

(*e*) The area bounded by the curve and the *x*-axis is unity. In other words, the area under the curve $= 1$.

(*f*) The probability that $x > x_A$ is the area to the right of *AA*.

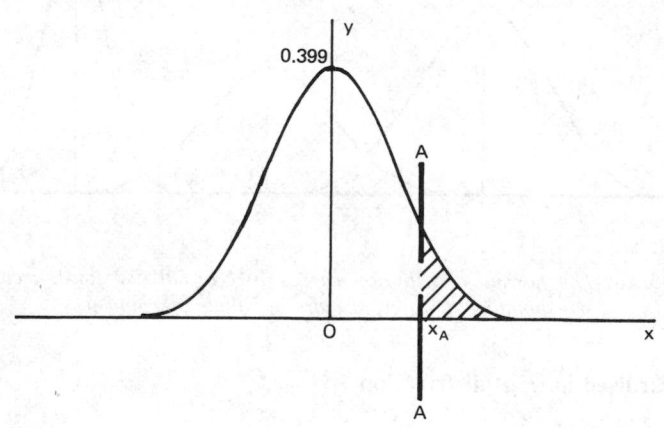

FIG. 96. *Standardised normal distribution curve.*

(*a*) Plot the standardised normal distribution curve on graph paper.

(*b*) By square counting calculate the area enclosed by the curve and the *x*-axis.

(*c*) By square counting calculate the proportion of the enclosed area lying to the right of the line $x = 1$ and hence probability that x lies between -1 and $+1$.

SOLUTION

(a) The plot is left as an exercise for the reader to calculate the x-y values for plotting by drawing up a suitable table e.g.

$$x: \quad 0 \quad \pm 0.2 \quad \pm 0.4 \quad \pm 0.6 \ldots \pm 3.6 \quad \pm 3.8$$
$$x^2: \quad 0 \quad 0.04 \quad 0.16 \quad 0.36 \ldots$$
$$e^{\frac{-x^2}{2}}: \quad 1 \ldots$$
$$y = \frac{1}{\sqrt{2\pi}} e^{\frac{-x}{2}}: \quad 0.399 \ldots$$

if cm graph paper is used suitable scales are:

$$x: 1 \text{ cm to } 1 \qquad y: 1 \text{ cm to } 0.1$$

if inch graph paper is used suitable scales are:

$$x: \tfrac{1}{2} \text{ inch to } 1 \qquad y: \tfrac{1}{2} \text{ inch to } 0.1$$

(b) Calculate the *area* scale thus:

cm graph paper: $1 \text{ cm}^2 = 1 \times 0.1 =$ area of 0.1 under curve
inch graph paper: $1 \text{ in}^2 = 2 \times 0.2 =$ area of 0.4 under curve

By square counting and applying the appropriate area scale, the area under the curve should be 1.00.

(c) By square counting, area to right of $x = 1$ is 0.16 approximately. By symmetry, area to left of $x = -1$ is 0.16 approximately

∴ area between $x = -1$ and $x = +1$ is approximately 0.68
∴ probability for $-1 < x < 1$ is 0.68 or 68% approximately.

Standard deviation

Standard deviation is the name given to the parameter, σ.
For the standardised curve, $\sigma = 1$ and standard deviation $= 1$.

(a) It is emphasised that *standard deviation* is a property of the theoretical curves which, strictly, apply only when there are an infinite number of observations. Mathematicians describe this as the standard deviation of the Universe or Population.

(b) It can be shown that for a very large number, n, of observations,

$$\sigma = \sqrt{\frac{\Sigma(u - \mu)^2}{n}}$$

where sigma (i.e. Σ) denotes "sum of".
$(u - \mu)$ is a measure of by how much each reading u differs from the mean, μ.

$(u - \mu)^2$ is the square of this quantity.

$\Sigma(u - \mu)^2$ is the sum of the squares of these deviations.

$\dfrac{(u - \mu)^2}{n}$ is the mean of the squares in n observations and $\sqrt{\dfrac{\Sigma(u - \mu)^2}{n}}$ is what physicists call the Root Mean Square (i.e. R.M.S.). Hence

$$\sigma = \text{standard deviation} = \text{R.M.S.}$$

(c) The results of a survey may involve the calculation of a value of the standard deviation for each of a number of different operations. For example, a base line may comprise a number of bays each of which has been measured several times but, perhaps, with different equipment for each bay. If the standard deviation for each bay is estimated to be $\sigma_1, \sigma_2, \sigma_3 \ldots$ then, for the whole base line the standard deviation would be:

$$\sigma = \sqrt{\sigma_1^2 + \sigma_2^2 + \sigma_3^2 + \ldots}$$

Weight

If the *weight* of each set of observations in the above example is denoted w_1, w_2, w_3 ... and vary from bay to bay, the over-all standard deviation will be

$$\sigma = \sqrt{w_1\sigma_1^2 + w_2\sigma^2 + w_3\sigma_3^2 + \ldots}$$

It should be noted that *weight* reflects the reliability of the measurement: that is, the greater the reliability the greater the weight.

Standard error

As indicated above σ is, strictly, a theoretical concept obtainable only from an infinite number of observations and so is μ. In practice, where only a limited number of observations can be made, the best that can be done is to make an estimate of σ. This is called the Standard Error (i.e. *se* or *s*). In the same way μ is a theoretical mean and the best estimate that can be made in practice is denoted $\bar{u}$ (i.e. "*u* bar"), the mean value of the readings in the sample. If n observations are made then, strictly, the standard error is defined as:

$$s = \sqrt{\dfrac{\Sigma(u - \bar{u})^2}{n - 1}}$$

Notice that $u - \bar{u}$ is the *residual*.

In surveying practice, however, the distinction between σ and s, or between μ and $\bar{u}$, is seldom made and standard deviation and standard error are often treated as being the same.

It should be noted that the formula above for s gives the standard error of a single observation. The standard error of the mean, s_m, is a different quantity but also obeys the rules governing normal distribution.

Hence
$$s_m \propto \frac{1}{\sqrt{n}}$$

Hence to reduce the standard error by 50% say, the number of observations has to be increased fourfold.

The difference in levels between two stations was determined by tacheometry, the staff station being higher than the instrument station. Afterwards, the standard error in V was estimated as ± 0.010 m, V being the difference in height between the staff mid-reading and the trunnion axis of the tacheometer. The height of collimation as measured was estimated to be correct ± 5 mm. If the mid-reading of the staff was approximately 1 metre and the verticality of the staff was correct within $\pm 6°$, determine:

(a) the accuracy in the calculated difference in level between the stations; and
(b) how many repeat observations would be necessary to bring the standard error to within 5 mm of the correct value.

SOLUTION
(a) Difference in level $= V + h - m$ (*see* Chapter 5)
$1 - \cos 6° = 0.0055$

hence standard error in mid-reading, $s_m = \pm 0.0055$
      "   "   " V $s_v = \pm 0.0100$
      "   "   " h $s_h = \pm 0.005$

hence, over all,
$$s = \sqrt{0.0055^2 + 0.01^2 + 0.005^2} = \pm 0.0125 \text{ m}$$
$$= \pm 12.5 \text{ mm}$$

(b) No. of observations required to reduce s to ± 5 mm is
$$\left(\frac{12.5}{5}\right)^2 = 6.25, \underline{\text{say } 7}$$

With reference to Specimen Question 54, the full expression for difference in level (*see* Chapter 5) is $\frac{1}{2} cs \sin 2\alpha + h - m$ and the overall standard error can therefore be calculated from the individual standard errors in c, s, α, h and m. To perform this calculation, however, a working knowledge of Partial Differentiation is necessary.

The more advanced techniques for distributing errors, which are described later in this chapter, also involve partial derivatives and a brief summary of the rules governing partial differentiation and its application is therefore given below.

PARTIAL DIFFERENTIATION

The mathematical statement $u = f(x)$ means "u is a function of x". If this equation is differentiated with respect to x, the expression obtained for $\dfrac{du}{dx}$ is called the derivative or differential coefficient and it is assumed that the reader knows the simple rules governing the processes by which these are obtained, i.e. the process of differentiation.

Complications arise when u is a function of more than one independent variable. It might, for instance, be a function of two variables, x and y, in which case the mathematician would write

$$u = f(x,y)$$

In this case, if y is treated as a constant (even though it is not), u can be differentiated with respect to x by the same rules which applied in the case of $y = f(x)$.

The result is not now written as $\dfrac{\partial u}{\partial x}$, however. Instead, to distinguish between the two different cases, it is written as $\dfrac{\partial u}{\partial x}$ the "curly dee" (i.e. ∂) being used to denote a partial derivative and this, in turn, implies that other variables (at least one) are being treated as though they were constants. In this case y is the variable which is being kept constant. Equally, x can be treated as a constant and u differentiated with respect to y, the result being written $\dfrac{\partial u}{\partial y}$. If u is a function of two variables, such as x and y above, there are two first partial derivatives $\dfrac{\partial u}{\partial x}$ and $\dfrac{\partial u}{\partial y}$.

For example, the area of a rectangle is given by $A = BD$. To find how A varies with (i) B and (ii) D, differentiate partially with respect to B and D in turn.

$$\frac{\partial A}{\partial B} = D \qquad \frac{\partial A}{\partial D} = B$$

Another example: the area of triangle ABC might be written

$$\triangle = \frac{c^2}{2} \frac{\sin A \sin B}{\sin C} \quad \text{using the usual symbols}$$

then $\qquad \dfrac{\partial \triangle}{\partial c} = \dfrac{c}{2} \dfrac{\sin A \sin B}{\sin C}$ $\quad A, B, C$ being treated as constants

$$\frac{\partial \triangle}{\partial A} = \frac{c^2}{2} \frac{\cos A \sin B}{\sin C} \quad c, B, C \qquad , \quad , \quad , \quad ,$$

$$\frac{\partial \triangle}{\partial B} = \frac{c^2}{2} \frac{\sin A \cos B}{\sin C} \quad c, A, C \qquad , \quad , \quad , \quad , .$$

The process can be extended to any number of independent variables. Hence, if $u = f(x, y, z\ldots\ldots)$ a partial derivative can be found for each independent variable and would be denoted $\dfrac{\partial u}{\partial x}, \dfrac{\partial u}{\partial y}, \dfrac{\partial u}{\partial z} \ldots$

Small changes
If $u = f(x,y)$, small changes δx, δy, in the independent variables produce a change δu in the dependent variable (i.e. u) which is given by

$$\delta u = \frac{\partial u}{\partial x}\delta x + \frac{\partial u}{\partial y}\delta y$$

Maxima and minima
If $u = f(x)$ a necessary condition for u to be a *maximum* or *minimum* is

that $\dfrac{\partial u}{\partial x} = 0$

if $u = f(x,y)$ the corresponding conditions for u to be a maximum or minimum are that both

$$\frac{\partial u}{\partial x} = 0 \quad \text{and} \quad \frac{\partial u}{\partial y} = 0$$

Similarly, if u is a function of, say, five independent variables there will be five such equations.

Principle of least squares
This principle states that if a number of observations are in error by amounts of $e_1, e_2, e_3 \ldots$ etc., the most probable result of the observa-

tions is that which makes the sum of the squares of the errors a minimum. Hence

$$\delta E = \Sigma e^2 = e_1{}^2 + e_2{}^2 + e_3{}^2 + \ldots$$

has to be a minimum.

i.e. $$\delta E = \frac{\partial E}{\partial e_1}\delta e_1 + \frac{\partial E}{\partial e_2}\delta e_1 + \frac{\partial E}{\partial e_3}\delta e_3 + \ldots = 0$$

when E is a minimum.

It should be noted that, just as the mathematical statement $u = f(x,y)$ means u is a function of x and y, so the statement $u = \phi(x,y)$ means u is a function of x and y.

The use of f and ϕ can imply that there are two different functions. Similarly any other letter can be used for the same purpose. Common ones are $u = F(x,y)$ $u = G(x,y)$ $u = H(x,y)$ etc.

Lagrange method of undetermined multipliers

The application of this technique is central to the Method of Correlates, *see* p. 206. If $u = \phi(x,y,z)$ and also $f(x,y,z) = 0$

Then $$\delta u = \frac{\partial \phi}{\partial x}\delta x + \frac{\partial \phi}{\partial y}\delta y + \frac{\partial \phi}{\partial z}\delta z$$

and, for a minimum, $\delta u = 0$

Hence $$\frac{\partial \phi}{\partial x}\delta x + \frac{\partial \phi}{\partial y}\delta y + \frac{\partial \phi}{\partial z}\delta z = 0 \qquad \text{(i)}$$

Also since $f(x,y,z) = 0$,

$$\frac{\partial f}{\partial x}\delta x + \frac{\partial f}{\partial y}\delta y + \frac{\partial f}{\partial z}\delta z = 0 \qquad \text{(ii)}$$

Multiply (ii) by an undetermined multiplier λ and subtract from (i):

$$\left(\frac{\partial \psi}{\partial x} - \lambda\frac{\partial f}{\partial x}\right)\delta x + \left(\frac{\partial \phi}{\partial y} - \lambda\frac{\partial f}{\partial y}\right)\delta y + \left(\frac{\partial \phi}{\partial z} - \lambda\frac{\partial f}{\partial z}\right)\delta z = 0$$

If λ is so chosen that the coefficient of δz, say, becomes zero then

$$\left(\frac{\partial \phi}{\partial x} - \lambda\frac{\partial f}{\partial x}\right)\delta x + \left(\frac{\partial \phi}{\partial y} - \lambda\frac{\partial f}{\partial y}\right)\delta y = 0$$

But since δx and δy are independent, their coefficients must also vanish. Hence:

$$\frac{\partial \phi}{\partial x} = \lambda\frac{\partial f}{\partial x}, \quad \frac{\partial \phi}{\partial y} = \lambda\frac{\partial f}{\partial y}, \quad \frac{\partial \phi}{\partial z} = \lambda\frac{\partial f}{\partial z}.$$

and the method can be extended to any number of variables.

It can also be extended to any number of functions. If, for example

$$u = \phi(x, y, z\ldots), f(x,y,z\ldots) = 0 \text{ and } F(x,y,z\ldots) = 0$$

then two multipliers λ and μ, say, are required and both can be evaluated. *See* Specimen Question 52 below and Specimen Question 55.

To illustrate how the Lagrange method operates Specimen Question 52 is now solved by this method.

SPECIMEN QUESTION 52 *see* p. 194.

SOLUTION

(*a*) *Equal Weights*

	Rise	Fall
AC	$132.877 + e_1$	
CD		$81.400 + e_2$
DA		$51.237 + e_3$

Since Σ Rises $= \Sigma$ Falls, $132.877 + e_1 = 132.637 + e_2 + e_3$

$$\therefore 0.240 = -e_1 + e_2 + e_3 \tag{i}$$

Also

$$\Sigma e^2 = e_1{}^2 + e_2{}^2 + e_3{}^2$$

has to be a minimum, u, say

$$\therefore \frac{\partial u}{\partial e_1}\delta e_1 + \frac{\partial u}{\partial e}\delta e_2 + \frac{\partial u}{\partial e_3}e_3 = 0$$

But

$$\frac{\delta u}{\delta e_1} = 2e_1, \frac{\partial u}{\partial e_2} = 2e_2, \frac{\partial u}{\partial e_3} = 2e_3$$

and, therefore,

$$e_1\delta e_1 + e_2\delta e_2 + e_3\delta e_3 = 0 \tag{ii}$$

Also, differentiating equation (i)

$$-\delta e_1 + \delta e_2 + \delta e_3 = 0 \tag{iii}$$

Multiplying (iii) by λ and subtract from (ii)

$$(e_1 + \lambda)\delta e_1 + (e_2 - \lambda)\delta e_2 + (e_3 - \lambda)\delta e_3 = 0$$

By Lagrange, all the coefficients must equal zero. Hence:

$$e_1 = -\lambda$$
$$e_2 = \lambda$$
$$e_3 = \lambda$$

Substitute these values in (i) to given

$$0.0240 = \lambda + \lambda - (-\lambda) = 3\lambda$$

$$\therefore \ \lambda = 0.08 = -e_1 = +e_2 = +e_3$$

$\therefore$ Corrected readings are

$$AC: +132.877 - 0.08 = 132.797$$
$$CD: -(81.400 + 0.08) = -81.480$$
$$DA: -(51.237 + 0.08) = -51.317$$
$$\text{Check: Closing Error} = 0$$

(b) *Weighted observations as on p. 195*

Because of the weighting of the observations, the condition is that $\Sigma we^2 = e_1{}^2 + 2e_2{}^2 + e_3{}^2$ has to be a minimum

$$\therefore \ e_1\delta e_1 + 2e_2\delta e_2 + e_3\delta e_3 = 0$$

as in (a)
$$-\delta e_1 + \delta e_2 + \delta e_3 = 0$$

Multiply by λ and subtract:

$$(e_1 + \lambda)\delta e_1 + (2e_2 - \lambda)\delta e_2 + (e_3 - \lambda)\delta e_3 = 0$$

and, since each coefficient = 0

$$e_1 = -\lambda$$
$$e_2 = \tfrac{1}{2}\lambda$$
$$e_3 = \lambda$$

but $0.240 = -e_1 + e_2 + e_3 = \lambda + \tfrac{1}{2}\lambda + \lambda = \dfrac{5}{2}\lambda$

$$\therefore \ \lambda = 2/5 \times 0.240 - 0.096$$

Hence the corrected readings become

$$AC: 132.781$$
$$CD: 81.448$$
$$DA: 51.333$$

METHOD OF CORRELATES

The solution immediately above is by the *method* of *correlates*. λ has been called the undetermined multiplier; in Surveying it is called a *correlate*. In Specimen Question 55 two correlates, λ and μ, have to be used and the simultaneous equations from which their values are determined are called the Condition Equations.

The procedure used in this method always follows the same pattern as the solution above to Specimen Question 52 (b), *see* p. 206. The steps are as follows.

(a) Assume the errors in the observations are e_1, e_2, e_3 ... with weights w_1, w_2, w_3....

(b) Write down the corrected value of the observation, thus:

first corrected value   = observed value $+ e_1$
second corrected value =   „   „ $+ e_2$
and so on.

(c) Sum the corrected values to give expressions for the closing errors in terms of e_1, e_2, e_3....

(d) Apply the principle of least squares, i.e. Σwe^2 has to be a minimum, and differentiate it partially to give: $w_1 e_1 \delta_1 e_1 + w_2 e_2 \delta_2 e_2 + \ldots = 0$.

(e) Differentiate partially the equations derived in (c) to give an equation of the form $\pm \delta e_1$, $\pm \delta e_2 \pm \ldots = 0$

(f) Multiply the equations derived in (e) by a correlate:

λ for the first, μ for the second, and so on.

Subtract each in turn from the equation derived in (d).

(g) Equate coefficients of δe_1, δe_2, ... to zero giving e_1, e_2 ... in terms of λ and μ and, if necessary, other correlates.

(h) Substitute the expressions from (g) into the equations from (c) to give simultaneous equations in λ, μ etc., from which λ, μ etc., can be determined.

(i) Evaluate e_1, e_2 ... from the expressions obtained in (g) and substitute these into the equations from (b) to give the corrected values of the observations.

SPECIMEN QUESTION 55

(a) Explain what is meant by *weight* as applied to an observed quantity.

(b) A, B, C and D are four angles closing the horizon at a triangulation station.

Observations were made as follows:

$$A = \quad 70° \ 14' \ 38'' \quad \text{(Weight 2)}$$
$$B = \quad 67° \ 20' \ 27'' \quad \text{(Weight 2)}$$
$$C = \quad 90° \ 01' \ 15'' \quad \text{(Weight 1)}$$
$$D = 132° \ 23' \ 31'' \quad \text{(Weight 1)}$$
$$(B + C) = 157° \ 21' \ 45'' \quad \text{(Weight 1)}$$

Determine the most probable values of the angles A, B, C and D.

SOLUTION

(a) *See* p. 195.

(b) *Steps* (a) *and* (b). Let errors be e_1, e_2, e_3, e_4, as indicated below.
Then corrected values are:

		Weight
A	$70° \; 14' \; 38'' + e_1$	2
B	$67° \; 20' \; 27'' + e_2$	2
C	$90° \; 01' \; 15'' + e_3$	1
D	$132° \; 23' \; 31'' + e_4$	1

Add: $359° \; 59' \; 51'' + e_1 + e_2 + e_3 + e_4$

			Weight
	$(B + C)$	$157° \; 21' \; 45'' + e_5$	1
Also	$B + C =$	$157° \; 21' \; 42'' + e_2 + e_3$	
Subtract		$03'' + e_5 - e_2 - e_3$	

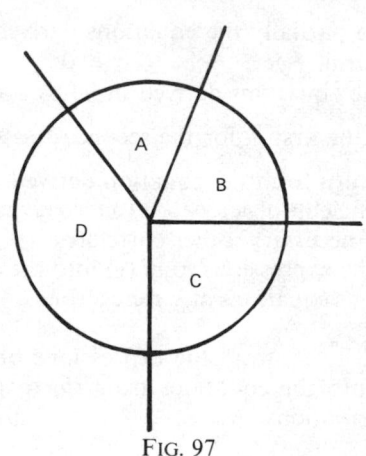

FIG. 97

Step (c)

$$\therefore \text{Closing error} = 09'' = e_1 + e_2 + e_3 + e_4 \qquad \text{I}$$
$$\text{and in } (B + C) \text{ error} = 03'' = \quad e_2 + e_3 \quad - e_5 \qquad \text{II}$$

Step (d)

$$\Sigma we^2 = 2e_1^2 + 2e_2^2 + e_3^2 + e_4^2 + e_5^2$$

and, differentiating,

$$2e_1\delta e_1 + 2e_2\delta e_2 + e_3\delta e_3 + e_4\delta e_4 + e_5\delta e_5 = 0 \qquad \text{III}$$

Step (e)

Differentiate equations I and II

$$\delta e_1 + \delta e_2 + \delta e_3 + \delta e_4 = 0 \qquad \text{IV}$$
$$\delta e_2 + \delta e_3 - \delta e_5 = 0 \qquad \text{V}$$

Step (f)

Multiply IV by λ, V by μ and subtract from III:

$$(2e_1 - \lambda)\delta e_1 + (2e_2 - \lambda - \mu)\delta e_2 + (e_3 - \lambda - \mu)\delta e_3 + (e_4 - \lambda)\delta e_4 + (e_5 + \mu)\delta e_5 = 0$$

Step (g)

Hence

$$e_1 = \tfrac{1}{2}\lambda$$
$$e_2 = \tfrac{1}{2}(\lambda + \mu)$$
$$e_3 = \lambda + \mu$$
$$e_4 = \lambda$$
$$e_5 = -\mu$$

Step (h)

Substitute for $e_1, e_2, e_3 \ldots$ in equations I and II:

$$\frac{\lambda}{2} + \frac{\lambda + \mu}{2} + (\lambda + \mu) + \lambda = 3\lambda + \frac{3}{2}\mu = 9$$

$$\frac{\lambda + \mu}{2} + (\lambda + \mu) + \mu = \frac{3}{2}\lambda + \frac{5}{2}\mu = 3$$

Solving simultaneously, $\quad \mu = -0.857$

$$\lambda = 3.428$$
$$\therefore \qquad \lambda + \mu = 2.57$$

Step (i)

Hence

$$e_1 = 1.7$$
$$e_2 = 1.3$$
$$e_3 = 2.6$$
$$e_4 = 3.4$$
$$e_5 = 0.9$$

$\therefore$ Most probable values are:

A:	70° 14′ 40″
B:	67° 20′ 28″
C:	90° 01′ 18″
D:	132° 23′ 34″

Check: Total $= 360°\ 00'\ 00''$

The subject of this example is a round of angles, the term used when the angles are measured right round a single point or station. Clearly, the condition to be satisfied is that after the corrections have been applied, the sum of the angles is 360° 00′ 00″. It is always worth making this check.

A round of levels, in which a level run follows a circuit and finishes at the starting point, presents what is mathematically the same problem. Either of these problems can be solved by the method of correlates; equally, either can be solved by one of the methods described below, which are a direct application of the Principle of Least Squares. The fundamental difference between these methods and the Method of Correlates occurs in the first step. In "Correlates" the initial assumption was that there would be errors $e_1, e_2, e_3 \ldots$ etc. In "Least Squares" the first step is to assume a value (unknown though it is) for the correct reading.

METHODS OF LEAST SQUARES

There are two different methods which differ only in the way the algebra is handled. The first method is more straightforward but its algebra can be very cumbersome. The second method is a refinement of the first, yields the same simultaneous equations and its algebra is less onerous; to use it, however, rules have to be memorised for dealing with the matrix. Both methods will be demonstrated by applying them to the problem in Specimen Question 56.

SPECIMEN QUESTION 56

(a) Levels were taken between four stations A, B, C and D with the following results in metres, $+$ indicating a rise;

A to B: $+62.110$ C to D: -81.400 D to A: -51.237
B to C: $+70.565$ A to C: 132.877

Taking the reduced level of A as datum and assuming all five observations to be of equal weight, calculate the most probable values for the reduced levels of B, C and D.

(b) What factors in practice might contribute to inequality in the weighting of the observations?

SOLUTION (FIRST METHOD)

Let B, C and D be at heights of b, c, d respectively above A. Then the residual errors are:

$$AB: \quad 62.110 - b$$
$$BC: \quad 70.565 - (c - b)$$
$$CD: \quad -81.400 - (d - c)$$
$$AC: \quad 132.877 - c$$
$$AD: \quad 51.237 - d$$

Note the reversal of DA to AD for the sake of consistency.

By the Principle of Least Squares, the minimum value of Σe^2 gives the most probable results. Also $E = \Sigma e^2$ is a minimum when

$$\frac{\partial E}{\partial b} = 0, \quad \frac{\partial E}{\partial c} = 0 \quad \text{and} \quad \frac{\partial E}{\partial d} = 0$$

Here, $E = (62.110 - b)^2 + (70.565 - c + b)^2 + (-81.400 - d + c)^2$
$$+ (132.877 - c)^2 + (51.237 -)^2$$

To differentiate, each term can be expanded. The algebra is simplified by treating each term as a function of a function. Using this technique:

$$\frac{\partial E}{\partial b} = 2(62.110 - b)(-1) + 2(70.565 - c + b)(+1)$$

$$\frac{\partial E}{\partial c} = 2(70.565 - c + b)(-1) + 2(-81.400 - d + c)(+1)$$
$$+ 2(132.877 - c)(-1)$$

$$\frac{\partial E}{\partial d} = 2(-81.400 - d + c)(-1) + (51.237 - d)(-1)$$

Dividing each of these equations throughout by 2 (a factor common to every term) collecting like terms and equating to zero yields the following simultaneous equations.

$$\begin{aligned} 2b - c + 8.455 &= 0 \\ -b + 3c - d - 284.842 &= 0 \\ -c + 2d + 30.163 &= 0 \end{aligned}$$

From which

$$b = 62.156$$
$$c = 132.766$$
$$d = 51.302$$

SOLUTION (SECOND METHOD)
This method starts exactly as the previous method:

Residuals:
$$\begin{aligned} AB: &\quad 62.110 - b \\ BC: &\quad 70.565 - (c - b) \\ CD: &\quad -81.400 - (d - c) \\ AC: &\quad 132.877 - c \\ AD: &\quad 51.237 - d \end{aligned}$$

Now, instead of squaring the residuals, differentiating and finding the three simultaneous equations given by $\frac{\partial E}{\partial b} = 0$, $\frac{\partial E}{\partial c} = 0$, and $\frac{\partial E}{\partial d} = 0$, a mechanical method is used which entails compiling a matrix, thus:

Weight	b	c	d	Observed value
1	−1	0	0	62.110
1	1	−1	0	70.565
1	0	1	−1	−81.400
1	0	−1	0	132.877
1	0	0	−1	51.237

(i) The matrix is constructed thus:

Weight column : enter weight of observed value.

"b" column : enter coefficient of b on each line where it exists.

"c" and "d" column : as for "b".

"Observed Value" : entered as it appears in the table of residuals.

(ii) To construct the first of the simultaneous equations the top two lines only are considered because, on the other three lines, the coefficient of b is zero. Each term in the top line is multiplied by the product of 1 (for the weight) and −1 (the coefficient of b) giving

$$(+1)(-1)\left(-b + 62.110\right)$$

Each term in the second line is multiplied by the product of 1 (weight) and +1 (coefficient of b) giving

$$(+1)(+1)(+b - c + 70.565)$$

These two are added to give the first of the simultaneous equations.

$$-(-b + 62.110) + (b - c + 70.565) = 0$$

i.e. $2b - c + 8.455 = 0$

(iii) To construct the second of the simultaneous equations, the procedure is the same, except that each term is multiplied by the product of the weight and the appropriate coefficient of c to give

$$(1)(-1)\left(b - c + 70.565\right) + (1)(1)\left(c - d - 81.400\right)$$

$$+ (1)(-1)\left(-c + 132.877\right) = 0$$

i.e. $-b + 3c - d - 284.842 = 0$

Three of the lines contribute to this equation because a coefficient of c appears in three lines.

Similarly, only two lines contribute to the third of the simultaneous equations, the procedure using the d coefficients being the same as for b and c above:

$$(1)(-1)\left(c - d - 81.400\right) + (1)(-1)\left(-d + 51.237\right) = 0$$

i.e. $\qquad\qquad -c + 2d + 30.163 = 0$

The three equations obtained are, therefore, the same as by the first method and, of course, lead to the same result which is given on p. 211.

NOTE

The reader should work through this example independently using both methods in order to assess which is easier. The explanations in the text make the second method appear protracted whereas the first method appears more compact than it really is.

If one or more of the observations is weighted, the weight in Method 2 becomes a multiplier of every term on its line, *see* Specimen Question 57.

SPECIMEN QUESTION 57

This is a modification of Specimen Question 56 with two observations weighted as indicated in the solution.

SOLUTION (FIRST METHOD)

Suppose the residuals and weights are now as follows:

		Weight
AB:	$61.110 - b$	1
BC:	$70.565 - (c - b)$	1
CD:	$-81.400 - (d - c)$	2
AC:	$132.877 - c$	3
AD:	$51.237 - d$	1

$$E = \Sigma we^2 = (62.110 - b)^2 + (70.565 - c + b)^2 + 2(-81.400 - d + c)^2 + 3(132.877 - c)^2 + (51.237 - d)^2$$

$$\frac{\partial E}{\partial b} = 2(-1)(62.110 - b) + 2(70.565 - c + b) = 0$$

$$\frac{\partial E}{\partial c} = 2(-1)(70.565 - c + b) + 2 \times 2(-81.400 - d + c) + 3 \times 2(-1)(132.877 - c) = 0$$

$$\frac{\partial E}{\partial d} = 2 \times 2(-1)(-81.400 - d + c) + 2(-1)(51.237 - d) = 0$$

Hence:

$$2b - c + 8.455 = 0$$
$$-b + 6c - 2d - 631.996 = 0$$
$$-2c + 3d + 111.563 = 0$$

giving:

$$b = 62.180$$
$$c = 132.814$$
$$d = 51.355$$

SOLUTION (SECOND METHOD)

From the residuals in the previous solution, set out the matrix thus:

Weight	b	c	d	Observed value
1	−1	0	0	62.110
1	+1	−1	0	70.565
2	0	1	−1	−81.400
3	0	−1	0	132.877
1	0	0	−1	51.237

Which leads to the same simultaneous equations and solution as Method 1.

Factors which might contribute to inequality in the weighting of the observations include the following:

(a) the observer and staffman;
(b) the equipment;
(c) the length of the sights;
(d) changing atmospheric conditions.

MORE COMPLICATED SURVEY ADJUSTMENTS

The reader should appreciate that problems occur in practice of a much more complex nature than those dealt with in this chapter, the aim of which has been to provide a sound basis for further reading.

EXAMINATION QUESTIONS

1. Repeat Specimen Question 55 using the methods of Least Squares.

2. Repeat Specimen Questions 56 and 57 using the method of correlates.

3. The horizontal length of a line PQ is known, within a metre, to be 500 m and is to be checked more accurately by subtense measurement. The subtense bar is 2 m long and is mounted horizontally and perpendicular to the line PQ. The theodolite which is to be used was tested separately and gave the following successive readings on a horizontal angle:

77° 04′ 15″, 16″, 14″, 14″, 13″, 16″, 17″, 15″, 18″, 14″, 13″, (the degrees and minutes remaining 77° 04′ for each repetition).

(a) From the data given determine the standard error of a single reading made with this instrument on a horizontal angle.

(b) Assuming the result obtained in (a) to apply to the subtense operation determine the number of readings required on the angle subtended by the targets of the subtense bar in order that the standard error in the length PQ shall not exceed 0.25 m. The standard error in the length of the subtense bar may be taken as negligible.

4. A triangle ABC, has the side AB measured as 1010 m, with a standard error of 0.5 m, angle A measured as 50° 00′ and angle B as 30° 00′. If the standard error in each of the angles is 1 minute, determine the standard error in the area of the triangle.

5. The difference of level between two stations is determined by tacheometry. The standard errors in the various quantities observed are:

In the vertical angle, θ	$\pm 30″$
In the staff intercept, s	± 5 mm
In the multiplying constant of the anallatic theodolite	± 0.002
In the reading of the middle hair	± 3.5 mm
In the height of the instrument axis above the station.	± 2 mm

If $\theta = +30°$, $s = 1500$ mm and the multiplying constant is 100, determine in mm the standard error in the calculated difference of level between the stations.

ANSWERS TO EXAMINATION QUESTIONS

Chapter 1
1. Collimation error, transverse axis and vertical axis.
2. Face-left and face-right readings.
3. Yes.
4. 11.4 m.
5. 0.060 m down in 30 m; correct by lowering diaphragm.
7. 12.29 m.

Chapter 2
1. (a) 148.44 m.
   (b) (i) 2°
       (ii) 3.5°
2. 270.381 m
3. 29.984 m.
5. 29.766 m.
6. 500.611 m.
8. 30.001 m.

Chapter 3
3. Significant R.L.s: $A = 80.94$ m; C.P.s = 77.55, 74.77 m.
   $\qquad\qquad\qquad B = 71.92$ m.
   Last O.B.M. 73.30 (i.e. with closing error = 0.005 m)
4. 0.020 m up in 70 m; true level at $B = 151.955$ m.
5. $Y_2 = \dfrac{11}{10}\left[(Y_2 - X_2) - (Y_1 - X_1)\right]$; adjustment of capstan screws
   which retain diaphragm.
6. Discrepancy between A + 30 and A + 40.
7. (b) 0.004 m
8. 1%.

Chapter 4
1. BC, 10 m short.
2. O: 545.1,   61.0
   G: 846.1, −830.7
   H: 121.6, −408.2
3. 3.5 ha.
4. A: 0   0
   B: 264.8 140.9
   C: 399.1   44.7
   D: 593.5 275.7
   E: 42.0 480.8
        E N
5. A 0   0
   B 103.90 124.48

   C 84.47   282.80
   D 62.99   318.72
6. (a) 3.01 m.
   (b) $\angle ABC$.
   (c) 0° 32′.
          E N
7. A 1842.18 2333.90
   B 1787.94 2199.71.
8. Mistake in length 216.6.
   True length = 236 approx.
   Co-ords. of K = (2192.7 E,
   1473.0 N).
9. (1617.1, 1469.2)

Chapter 5

2. 101; 0.3 m.
3. 16.5%—say 1 in 6.
4. (i) 216.460.
   (ii) 189.704.
5. (a) 48.756 m.
   (b) B: 55.427 A.O.D.
       C: 65.806 A.O.D.
6. (a) 69.468 m.
       62.760 m.
   (b) 106.258 A.O.D.
       98.172 A.O.D.
7. XY: 261.66.

YZ: 256.85.
XZ: 238.45.
height diffcrence = 26.488 m.
9. $ODH = 79° 47' 22''$
   $DHO = 58° 32' 40''$
   $HOD = 41° 39' 58''$
   $OHG = 77° 29' 02''$
   $HGO = 36° 02' 18''$
   $GOH = 66° 28' 40''$
   $OGC = 88° 22' 10''$
   $GCO = 71° 29' 28''$
   $COG = 26° 08' 22''$

Chapter 6

1. (a) (i) 86.60 m.
       (ii) 157.08m.
       (iii) 60°.
2. $R = 290$ m.
   $L = 603.6$ m.
3. T.P. at X; by 2.0 m.
4. 11 921 m; 13 589 m.
5. 1111 m radius

Chainage	S. O. Angle
50	358° 42' 40''
100	357° 25' 20''
150	356° 08' 00''
200	354° 50' 40''
250	353° 33' 20''
300	352° 16' 00''
350	350° 58' 20''
382	350° 09' 00''

6. 86.535 m; 0.061 m in 600 m, up.
7. (b) 274.59 m.
8. 0.671 m.
9. 18.275.
   0.074 "down" in 594 m
10. $\angle PAB$: 63° 45' 01''
    $\angle PBA$: 70° 46' 35''
    $\angle BCP$: 60° 54' 39''.
11. 800 m from P $\Big\}$ by graphical solution
    640 from R
12. 353 m.

Chapter 7

1. | Chainage | Reduced level |
       |----------|---------------|
       | 5980 | 124.800 |
       | 6010 | 125.074 |
       | 6040 | 125.297 |
       | 6070 | 125.467 |
       | 6100 | 125.586 |
       | 6130 | 125.635 |
       | 6160 | 125.633 |
       | 6190 | 125.578 |
       | 6220 | 125.472 |

2. | 3280 | 197.500 |
       |------|---------|
       | 3310 | 197.047 |
       | 3340 | 196.687 |
       | 3370 | 196.422 |
       | 3400 | 196.250 |
       | 3430 | 196.172 |
       | 3460 | 196.187 |
       | 3490 | 196.297 |
       | 3520 | 196.500 |

Lowest point: 3440 m, 196.167 A.O.D.

4. (a)   73.40 A.O.D.
         850.00 m.
   (b) 4%.
   (c) 77.994 A.O.D.
   (d) 254.67 m.

5. (a)   119.76 m.
   (b) 3023.60 m.
         20.096 A.O.D.
   (c) 3103.44 m.
         21.826 A.O.D.

6. (a) $R = 1448.5$ m, $L = 1252.51$ m
   (b) | x | δ |
        |-----|----------|
        | 0 | $00°\ 00'\ 00''$ |
        | 15 | $00°\ 00'\ 39''$ |
        | 30 | $02'\ 35''$ |
        | 45 | $05'\ 48''$ |

7. (a) 1140.5 m.
   (b)   62.64 m
   (c) | x | δ |
        |-----|----------|
        | 0 | $00'\ 00''$ |
        | 10 | $00'\ 48''$ |
        | 20 | $03'\ 17''$ |
        | 30 | $07'\ 13''$ |
        | 40 | $12'\ 50''$ |

50	20′ 03″
62.64	31′ 28″

(*d*) 4116.73.

8. 1100 m; 1190 m.
   05′ 42″, 22′ 54″, 51′ 36″; 0.138 m.

9. (*i*) 1485.74 m; (*ii*) Set up at *B* with reading 3° 16′ 31″.

   (*iii*) Chainage   Deflection angle

Chainage	Deflection angle
1500	00° 49′ 00″
1515	01° 40′ 40″
1530	02° 32′ 00″

10. (*a*) 245 m; (*b*) 480 m; (*c*) Chainage (*offset to nearest 5 mm*)

Chainage	offset
20 m	25 mm
40	220
61.2	775

Chapter 8

1. 1.2670 ha.
2. 598.9 m.
3. 195 mm.
4. (*a*) 48° 06′ 31″.
   (*b*) 66° 30′ 10″.

Chapter 9

1. (Horizontal): £237 000 x
   (Inclined): £462 000 x.
2. 1.663 hectares.
3. (*a*) 61 000 m³.
   (*b*) 2 220 000 m³.
4. 18.3 m.
   10.2 m.
5. (*a*) 3.68 × 10⁶ m³.
   (*b*) 1900 m³.
6. 94 600 m³.
7. 21 703; 21 708 m³.
8. 22 500 m³.
9. 148 000 m³.
10. 160 400 m³.

Chapter 10

1. 37° 53′ 46″ S; 80° 32′ 14″; S; 32° 07′ 55″ N at upper transit
           13° 15′ 00″ S at lower transit.
2. N 52° 19′ 48″.

Chapter 11

4. (b) 22; 2400; 96 m/s.
5. (c) 120.6 m.
7. 38 m to be deducted.

Chapter 12

3. (a) 1.732″.
   (b) 18.
4. Area = 198 374 m² ± 226 m².
5. 223 mm.

INDEX

GEORGE GODWIN LIMITED
The book publishing subsidiary
of the Builder Group

Details of other **Godwin Study Guides**
can be found on the following pages.

For a full list of other George Godwin books
on related subjects, write for the FREE
Construction Book Catalogue available from:

George Godwin Ltd
1—3 Pemberton Row
Red Lion Court
Fleet Street
London EC4

Godwin Study Guides
General editor: M.J. SMITH
This series aims to provide the student preparing for TEC Higher National, B.Sc. and professional examinations with a clear concise guide to the principles of his subject. Basic theory is reinforced by means of graded worked examples carefully selected to illustrate the text. The books will also be of value to professional engineers in mid-career who need to refresh or update their grasp of specialist subjects. In view of this wider readership, and the fact that many students may require a greater coverage of a particular subject, books in the series have now been expanded to cover a wider range of topics, while maintaining the concise form of presentation. SI units are used throughout.
Titles in the series to date are as follows:

Materials and Structures
M.J. SMITH
This book is an elementary text on strength of materials and theory of structures for civil and structural engineers. It also meets the requirements in this subject of architects, quantity surveyors and students in other branches of engineering. The level of the text is especially appropriate for first-year students preparing for TEC Higher Diplomas and Certificates and B.Sc. and C.E.I. examinations. The book may also be taken as an introduction to the companion volumes, *Theory of Structures* and *Advanced Theory of Structures.*
Illustrated

Theory of Structures
M.J. SMITH & BRIAN J. BELL
Prepared as the next step in the study of theory of structures following *Materials and Structures,* this book is set at an intermediate standard for civil and structural engineers and deals with problems that cannot be solved by simple statics, owing to redundant forces. The latest edition has been fully revised and updated.
Illustrated

Advanced Theory of Structures
BRIAN J. BELL
Designed as the sequel to *Theory of Structures* this volume
tackles problems involving indeterminacy, such as continuous
member, trussed frame or portal frame. The book is considered
appropriate for students preparing for a university or C.N.A.A.
degree in Structural or Civil Engineering, for C.E.I. or Society
of Engineers examinations and for TEC Higher Certificate
and Higher Diploma courses.
Illustrated

Design of Reinforced Concrete Elements
R.W. CLEMENTS
The purpose of this book is to give guidance in the design of
simple reinforced concrete elements and to assist students of
building and civil and structural engineering to second-year
degree level or for TEC Higher Certificate and Diploma courses.
Design examples are based on the limit state concept, the
explanation of concepts inherent in limit state design being
kept as simple as possible. Worked examples have been prepared in SI units.
Illustrated

Structural Steelwork
F.W. LAMBERT
This book deals with the design of structural steelwork, con-
centrating on the elements of the structure rather than the
complete framework. Due attention is paid throughout to
welding and bolting for both sub-assembly in the workshop
and erection at site. In this revised edition Chapters 1 and 2
have been altered and extended to take account of the amend-
ment to B.S. 449.
Illustrated

Specifications and Quantities
DAVID BURCHESS
This is a useful introduction to the task of preparing specifications and bills of quantities for engineering projects. The latest edition has been updated, in particular to take account of revisions in the Conditions of Contract, and the publication of the Civil Engineering Standard Method of Measurement.
Illustrated

Soil Mechanics
M.J. SMITH
This book concentrates attention on basic principles. The first six chapters deal with soil composition, compaction, classification, permeability and flow nets, consolidation and shear strength. The final three chapters introduce a practical approach to the analysis of pressure on retaining walls, stability of slopes and foundation design.
Illustrated

Fluid Mechanics
R.H. DUGDALE & W.S. BANNISTER
This book covers more than merely the civil engineering aspects of fluid mechanics: in particular, the authors show how widely the principles of conservation of energy and conservation of momentum can be applied.
Illustrated

Mathematics for Engineers
S.F. HANCOCK
The author of this book has aimed throughout at developing the reader's practical mathematical ability. The simpler mathematical processes have been omitted, but the more advanced ones are illustrated profusely with many worked specimen questions, and each chapter ends with a selection of recent examination questions.
Illustrated

9082